Graph Theory

WILEY-INTERSCIENCE
SERIES IN DISCRETE MATHEMATICS AND OPTIMIZATION

ADVISORY EDITORS

RONALD L. GRAHAM
AT & T Laboratories, Florham Park, New Jersey, U.S.A.

JAN KAREL LENSTRA
*Department of Mathematics and Computer Science,
Eindhoven University of Technology, Eindhoven, The Netherlands*

JOEL H. SPENCER
Courant Institute, New York, New York, U.S.A.

A complete list of titles in this series appears at the end of this volume.

Graph Theory

RUSSELL MERRIS
California State University, Hayward

A Wiley-Interscience Publication
JOHN WILEY & SONS, INC.
New York • Chichester • Weinheim • Brisbane • Singapore • Toronto

This book is printed on acid-free paper. ∞

Copyright © 2001 by John Wiley & Sons, Inc. All rights reserved.

Published simultaneously in Canada.

No part of this publication may be reproduced, stored in a retrieval system or transmitted in any form or by any means, electronic, mechanical, photocopying, recording, scanning or otherwise, except as permitted under Section 107 or 108 of the 1976 United States Copyright Act, without either the prior written permission of the Publisher, or authorization through payment of the appropriate per-copy fee to the Copyright Clearance Center, 222 Rosewood Drive, Danvers, MA 01923, (978) 750-8400, fax (978) 750-4744. Requests to the Publisher for permission should be addressed to the Permissions Department, John Wiley & Sons, Inc., 605 Third Avenue, New York, NY 10158-0012, (212) 850-6011, fax (212) 850-6008, E-mail: PERMREQ@WILEY.COM.

For ordering and customer service, call 1-800-CALL-WILEY.

Library of Congress Cataloging-in-Publication Data:

Merris, Russell, 1943-
  Graph theory / Russell Merris.
    p. cm. — (Wiley-Interscience series in discrete mathematics and optimization)
  "A Wiley-Interscience publication."
  Includes bibliographical references and index.
  ISBN 0-471-38925-0 (cloth : alk. paper)
  1. Graph Theory.   I. Title.   II. Series.
  QA166.M45 2000
  511'.5—dc21
                                                                00-036785

Printed in the United States of America.

10 9 8 7 6 5 4 3 2 1

*To
Kenneth R. Rebman,
friend and mentor*

Contents

Preface   ix

1 Invariants   1

Definitions; pros and cons of pictures; different versus nonisomorphic graphs; isomorphism problem, invariants; "first theorem of graph theory"; adjacency matrix.

2 Chromatic Number   21

Proper colorings; Fundamental Counting Principle; independence, clique, and chromatic numbers; chromatic polynomial; unions and joins; bipartite graphs, trees; paraffins, Wiener and Balaban (chemical) indices.

3 Connectivity   45

Quantitative measures of connectivity; blocks, k-connectedness; separating sets, internally disjoint paths, Menger's Theorem; Whitney's Broken Cycle Theorem.

4 Planar Graphs   63

Euler's Formula; Kuratowski's Theorem; Five and Four-Color Theorems; geometric dual; embeddings in orientable surfaces, graph genus; theorems of Heawood and of Ringel and Youngs.

5 Hamiltonian Cycles   83

Necessary conditions; sufficient conditions; closure; Chvátal graphs; hamiltonian plane graphs; theorems of Whitney and Grinberg.

6 Matchings   103

Kekulé and benzene; perfect matchings; matching polynomial; adjacency characteristic polynomial; matching and covering numbers, Egerváry–König Theorem from Menger's Theorem; Theorems of Hall and Tutte.

7 Graphic Sequences 125

Partitions, graphic partitions, Havel–Hakimi criterion; graphs with the same degree sequence, Ryser switches; majorization, Ferrers diagrams; Ruch–Gutman criterion; threshold partitions and graphs.

8 Chordal Graphs 147

Weak majorization, shifted shapes, Ruch–Gutman criterion revealed; split partitions and graphs; chordal graphs; perfect graphs; simplicial vertices and chromatic polynomials.

9 Oriented Graphs 171

Acyclic orientations, Stanley's Theorem; Laplacian matrix, spanning tree number; spectral characterization of certain graph invariants; isospectral and decomposable graphs.

10 Edge Colorings 195

Ramsey numbers, upper and lower bounds, Erdös's probabilistic technique; Pólya-Redfield approach to graph enumeration, cycle index polynomial.

Hints and Answers to Selected Odd-Numbered Exercises   211

Bibliography   227

Index   229

Index of Notation   235

Preface

Where it can be found at all, an undergraduate course in graph theory is typically listed among the electives for majors in the mathematical sciences. For better or worse, elective courses often find themselves competing with each other for enrollments in today's cafeteria-style curriculum. Evolving in such an environment, this book might best be described as an *invitation* to graph theory. A mathematically rigorous introduction, it is neither an exhaustive overview nor an encyclopedic reference. To borrow a phrase from the early calculus reform movement, it is a *lean and lively* text, designed to attract and engage. Preference is given to the kinds of things students can, figuratively speaking, get their hands on. The selection is dictated less by intrinsic importance and more by how well a topic lends itself to being pictured, calculated, manipulated, or counted. While applications are numerous, they have been chosen and presented as much to seduce as to inform. In a similar spirit, not because it is central to graph theory, but because it is a useful device, graph coloring is deployed throughout the book as a unifying theme. The idea is to create the impression, not of a series of theorems about graphs, but of a subject that can be represented as a unified theory. The result is a book that goes deeply enough into a sufficiently wide variety of topics to display, not only the flavor of graph theory, but some of its power and elegance as well.

The text was planned to be a versatile instructional tool. Following a basic foundation in Chapters 1–3, an instructor is free to pick and choose the most appropriate topics from four independent strands. (Occasional redundancy is the price for such flexibility; for example, some definitions can be found in more than one place.) Specific dependencies among the chapters are illustrated by the diagram.

Beyond the (in)dependence indicated in the chart, Chapter 5 does not rely on any material coming after Example 4.14; embeddings of graphs in orientable surfaces of positive genus can be safely omitted. Moreover, the material on Ramsey Theory in Chapter 10 can be covered independently of the application of Pólya's Theorem to graph enumeration that concludes the book.

The prerequisites for a course based on this book should include the sophomore linear algebra course commonly found among the lower division requirements for majors in the mathematical and physical sciences. Elementary linear algebra is used most heavily in Chapter 9, where the reader is presumed to have been

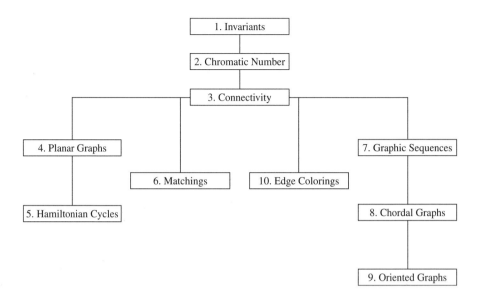

exposed to the definitions of eigenvalue, eigenvector, and the characteristic polynomial of a real matrix. Apart from general prerequisites, the recipe for counting nonisomorphic graphs, found in the latter part of Chapter 10, supposes the reader to be acquainted with the disjoint cycle factorization of a permutation.

Confounding any discussion of graph connectivity is an awkward discrepancy between intuition and proof. For this reason, Chapter 3 is likely to be the most difficult section of the text. In approaching this material, it is perfectly acceptable for an instructor simply to state some of the deeper results and *glide nimbly* over the details of proof.

Many people have contributed observations, suggestions, and constructive criticisms at various stages of production. Among them are Rob Beezer of the University of Puget Sound, Richard Brualdi of the University of Wisconsin, Ron Graham of UC San Diego, Jerry Griggs of the University of South Carolina, Matt Hubbard and Tom Roby of CSU Hayward, Ken Rebman of the University of Colorado, Colorado Springs, Peter Rowlinson of the University of Stirling, Edward Scheinerman of Johns Hopkins University, Linda Valdes of San José State University, and Steve Quigley of John Wiley & Sons. I would also like to acknowledge helpful conversations with Carl Johnson and Rhian Merris. Finally, I am deeply grateful to Dean Michael Leung of CSU Hayward, who, at a critical point of the project, granted me an extra portion of that most precious commodity — time.

Despite everyone's best efforts, no book seems complete without a few errors. An up-to-date errata, accessible from the internet, will be maintained at URL

http://www.telecom.csuhayward.edu/~rmerris

Preface

Appropriate acknowledgment will be extended to the first person who communicates the specifics of each previously unlisted error, preferably by e-mail addressed to merris@csuhayward.edu.

RUSSELL MERRIS

Hayward, California
July, 2000

Graph Theory

1

Invariants

By the end of a long flight, most airline passengers will have become familiar with the contents of the seat pocket in front of them. Perhaps the most interesting thing to be stowed there is the airline's route map. On the standard map, direct service between cities A and B is indicated by an arc linking the two. Some of these arcs may cross at points where airliners do not touch down, and it is the job of air traffic controllers to ensure that nothing unpleasant occurs at such places. This is usually accomplished by requiring planes on intersecting routes to fly at different altitudes. Thus, while arcs representing routes may cross, the routes themselves meet only at airports.

The graph-theoretic name for the number of arcs touching at city C is the *degree* of C. Would you be surprised to learn that, on every airline route map, at least one pair of cities must share the same degree?

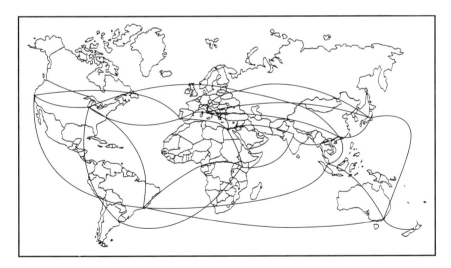

Some cities may be illustrated on the map even though they are not serviced by the airline. These have degree 0. It may be, on the other hand, that one or more of the indicated cities is an airline "hub" from which passengers may take direct flights to every other city on the map. It is easy to see, however, that these two extreme cases cannot both occur on the same map. While the degrees of the k cities range between 0 and $k - 1$, if there are arcs from one of them to all the others, then no city can have degree 0. Thus, among the k cities on any given map, at most $k - 1$ degrees are possible. Because there are more cities than degrees, some two cities must have the same degree.

Airline route maps provide just one representation of the mathematical abstraction called a *graph*. Informally, a graph is a finite collection of dots, some pairs of which are joined by arcs. In order to facilitate our discussion of these objects, let's agree on a few standard examples. The *path*, P_n, consists of $n \geq 2$ dots, representing *vertices*, and $n - 1$ arcs, representing *edges*. Two vertices of P_n have degree 1, and the rest are of degree 2.

As if its vertices were stepping stones, the *length* of P_n is $n - 1$, the number of steps from one degree 1 vertex to the other. The first few paths are illustrated in Fig. 1.1.

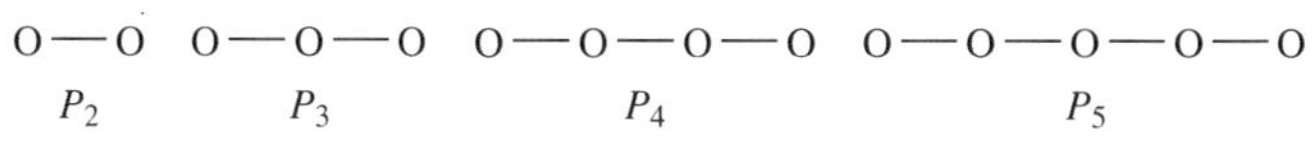

Figure 1.1. Paths.

The *star*, S_n, has $n \geq 2$ vertices and $n - 1$ edges. As shown in Fig. 1.2, $S_2 = P_2$ and $S_3 = P_3$. For $n \geq 3$, S_n has a central vertex of degree $n - 1$ and $n - 1$ vertices of degree 1.

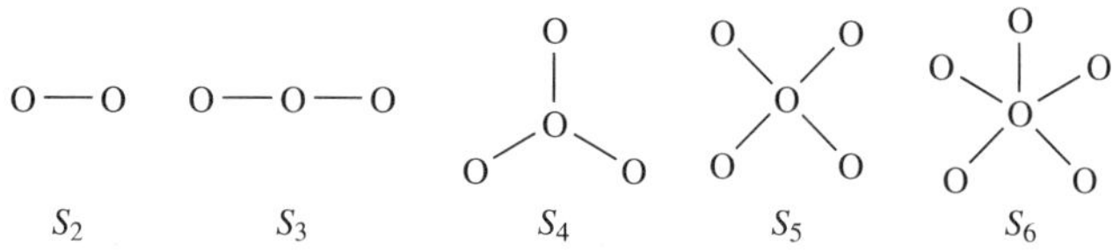

Figure 1.2. Stars.

The first few *cycles* are illustrated in Fig. 1.3. With n vertices and n edges the cycle, C_n, has length n.

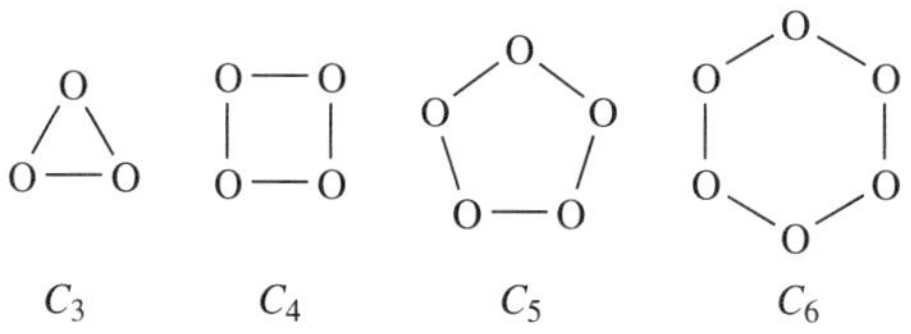

Figure 1.3. Cycles.

While a picture may be worth a thousand words, informal discussions based on pictures have serious limitations. Consider, for example, Fig. 1.4. Does it

Chap. 1 Invariants

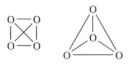

Figure 1.4

illustrate two different graphs, two different pictures of the same graph, or a single graph with two connected components? Before we can hope to give a definitive answer, much less prove any theorems, we need a rigorous (mathematical) definition of "graph." As a first step in that direction, consider the set $V = \{a, b, c, d, e\}$. Among the $2^5 = 32$ subsets of V, only one (the empty set) has no elements. There are five one-element subsets and $C(5, 2) = 10$ two-element subsets,[1] namely,

$$\begin{array}{llll} \{a,b\} & \{a,c\} & \{a,d\} & \{a,e\} \\ & \{b,c\} & \{b,d\} & \{b,e\} \\ & & \{c,d\} & \{c,e\} \\ & & & \{d,e\} \end{array}$$

It is these two-element subsets that are of interest. If V is an n-element set, denote by $V^{(2)}$ the set consisting of all $C(n, 2)$ of its two-element subsets.

1.1 Definition. A *graph* consists of two things: a nonempty finite set[2] V and a (possibly empty) subset E of $V^{(2)}$. Typically written $G = (V, E)$, the elements of V are the *vertices* of G, and the elements of E are its *edges*. When more than one graph is under consideration, it may be useful to write $V(G)$ and $E(G)$ for its vertex and edge sets, respectively.

An airline route map shows a graph illustrated on top of an ordinary map. The graph is drawn so that its vertices correspond to cities while its edges correspond to arcs. In the route map context, arc length is a rough-and-ready guide to the length of the corresponding flight, whether measured in travel time or frequent flyer miles. This metric property is not reflective of the graph. The length of a representing arc has no graph-theoretic meaning. Indeed, while an arc contains infinitely many geometric points, the edge it represents consists of (just) two vertices.

1.2 Definition. If $e = \{u, v\} \in E(G)$, vertices u and v are said to be *adjacent* (to each other) and *incident* to e. Two edges are *adjacent* if they have exactly one common vertex—that is, if their set-theoretic intersection has cardinality 1.

---

[1] Read "n-choose-r," binomial coefficient $C(n, r) = n!/[r!(n-r)!]$.
[2] Infinite graphs are not discussed in this book.

1.3 Example. Suppose $V = \{1, 2, 3, 4\}$ and $E = V^{(2)}$. Then $G = (V, E)$ is a *complete* graph. (One interpretation of Fig. 1.4 is that it contains two illustrations of $G = (V, E)$.)

If $W = \{1, 2, 3, 4, 5\}$, then $W^{(2)}$ has $C(5, 2) = 10$ elements. Thus, $W^{(2)}$ has 2^{10} subsets. Evidently, there are 1024 different graphs with vertex set W. One of them is $H = (W, F)$, where $F = \{\{1, 3\}, \{2, 3\}, \{2, 5\}, \{3, 4\}, \{3, 5\}, \{4, 5\}\}$. (The notation indicates that $V(H) = W$ and $E(H) = F$.) Two illustrations of H appear in Fig. 1.5. □

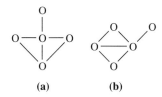

(a)   (b)

Figure 1.5

1.4 Definition. Let $G = (V, E)$ be a graph. Suppose that $v \in V$. The *degree* of v, denoted $d_G(v)$, is the number of edges of G that are incident with it. Alternatively, $d_G(v)$ is the number of vertices of G that are adjacent to v. When its meaning is clear, we will sometimes write $d(v)$ in place of $d_G(v)$. Let $\Delta(G)$ and $\delta(G)$ be the largest and smallest of the vertex degrees of G, respectively.

Given a graph G with $V(G) = \{v_1, v_2, \ldots, v_n\}$, it is useful to view the multiset $\{d(v_1), d(v_2), \ldots, d(v_n)\}$ as a sequence. Let $d(G) = (d_1, d_2, \ldots, d_n)$, where $\Delta(G) = d_1 \geq d_2 \geq \cdots \geq d_n = \delta(G)$ are the degrees of the vertices of G (re)arranged in nonincreasing order. (Note that d_i need not equal $d(v_i)$.) For example, $d(P_n) = (2, 2, \ldots, 2, 1, 1)$, $d(S_n) = (n - 1, 1, \ldots, 1)$, and $d(C_n) = (2, 2, \ldots, 2)$. Observe that

$$\Delta(G) \leq n - 1,$$

for any graph on n vertices, with equality if and only if G has a *dominating* vertex, one that is adjacent to every other vertex of G.

Let H be the graph from Example 1.3, illustrated twice in Fig. 1.5, Then $d(1) = 1 = \delta(H)$, $d(2) = 2$, $d(3) = 4 = \Delta(H)$, $d(4) = 2$, $d(5) = 3$, and $d(H) = (4, 3, 2, 2, 1)$. (Note that, sure enough, two vertices share the same degree!) From their degrees, it is an easy matter to identify the dots corresponding to vertices 1, 3, and 5 in Fig. 1.5. But, which of the remaining dots represents vertex 2 and which of them corresponds to vertex 4? In this case, either way of assigning dots to the even-numbered vertices results in a correct illustration.

1.5 Notation. The shorthand notation $uv \in E$ is sometimes used to indicate that $\{u, v\}$ is an edge of $G = (V, E)$.

Chap. 1 Invariants

1.6 Example. While different pictures can illustrate the same graph, it is also the case that different graphs can be illustrated by the same picture! If $V = \{a, b, c, d, e\}$ and $E = \{ac, bc, be, cd, ce, de\}$, then $G = (V, E)$ is illustrated by the pictures in Fig. 1.5. □

If the truth be known, "different" graphs are much less interesting than "nonisomorphic" graphs. Roughly speaking, graphs that can be illustrated by the same picture are said to be isomorphic. Let's make this description more precise.

1.7 Definition. Graph $G_1 = (V_1, E_1)$ is *isomorphic* to $G_2 = (V_2, E_2)$ if there is a one-to-one function f from V_1 onto V_2 such that $\{u, v\} \in E_1$ if and only if $\{f(u), f(v)\} \in E_2$. If such a function exists, it is called an *isomorphism* from G_1 to G_2.

From Exercise 4 (below), $f: V(G_1) \to V(G_2)$ is an isomorphism from G_1 to G_2 if and only if $f^{-1}: V(G_2) \to V(G_1)$ is an isomorphism from G_2 to G_1. In other words, G_1 is isomorphic to G_2 if and only if G_2 is isomorphic to G_1. We will write $G_1 \cong G_2$ to indicate that the graphs are isomorphic (to each other). Evidently, isomorphic graphs have the same numbers of vertices and edges. Properties like these, which isomorphic graphs must share, are called *invariants*.

1.8 Example. Let H and G be the graphs in Examples 1.3 and 1.6, respectively. Then $f(1) = a$, $f(2) = b$, $f(3) = c$, $f(4) = d$, and $f(5) = e$ is an isomorphism from H to G. It is not the only one; $g(1) = a$, $g(2) = d$, $g(3) = c$, $g(4) = b$, and $g(5) = e$ is another. Indeed, $h(1) = 1$, $h(2) = 4$, $h(3) = 3$, $h(4) = 2$, and $h(5) = 5$ is an isomorphism from H to itself.[3] This indicates why it did not matter which of the two dots of degree 2 in Fig. 1.5 is assigned to vertex 2 and which to vertex 4. □

1.9 Theorem. *Isomorphic graphs have the same degree sequence, that is, the degree sequence is an invariant.*

Proof. Let G and H be isomorphic graphs. If $u \in V(G)$, then $d_G(u) = o(\{v : uv \in E(G)\})$, the cardinality of the set of vertices adjacent to u. Suppose $f: V(G) \to V(H)$ is an isomorphism. Then f is one-to-one and onto, and $uv \in E(G)$ if and only if $f(u)f(v) \in E(H)$. It follows that

$$d_G(u) = o(\{f(v) : f(u)f(v) \in E(H)\})$$
$$= d_H(f(u)).$$

Thus, f induces a one-to-one correspondence between the multiset of vertex degrees of G and the multiset of vertex degrees of H. ∎

---
[3] An isomorphism from H to itself is called an *automorphism* of H.

The proof of Theorem 1.9 shows that isomorphisms preserve degrees vertex by vertex. In other words, no isomorphism can carry a vertex of degree 5 to a vertex of degree 4 or to one of degree 6.

1.10 Example. Among the standard examples in graph theory is the *Petersen graph*, shown twice in Fig. 1.6. The proof that G_1 and G_2 are isomorphic is "by the numbers." Let $V(G_1) = \{1, 2, \ldots, 10\} = V(G_2)$, as illustrated in Fig. 1.6. Define $f: V(G_1) \to V(G_2)$ by $f(i) = i$, $1 \leq i \leq 10$. Then, as may be verified by carefully examining the pictures, f is an isomorphism. (Check it out. Confirm that vertices i and j are adjacent in G_1 if and only if they are adjacent in G_2.) Provided that the numbers "check out," such a pair of labeled figures constitutes a proof of isomorphism.

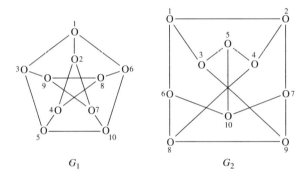

Figure 1.6. Two illustrations of the Petersen graph.

Note that, while the arc (segment) representing edge $\{2, 4\}$ crosses the arc corresponding to $\{8, 9\}$ in G_1, their point of intersection does not represent a vertex: $\{8, 9\} \cap \{2, 4\} = \phi$. □

Confirming that the vertex numberings in Fig. 1.6 correspond to an isomorphism is not the same thing as finding the numberings in the first place. But, how hard can it be? Evidently, it suffices to number the vertices of G_1 in some fixed but arbitrary way and then check to see whether any of the 10! numberings of the vertices of G_2 produces an isomorphism. (Because all the vertices have degree 3, there aren't any obvious shortcuts). If it takes a minute to check each numbering, then by working 24 hours a day the 10! = 3,628,800 possibilities can be checked out in just a little under 7 years. (If the graphs has a dozen vertices, the job would take 700 years!) Of course, this calculation assumes that no isomorphism is found before checking the last numbering of G_2. It seems that the worst case is when the graphs are *not* isomorphic, precisely the case in which invariants are most useful. If some invariant is not the same for a pair of graphs, there is no need to check numberings. The graphs cannot be isomorphic!

Chap. 1 Invariants

The search for a short list of easily computed invariants sufficient to distinguish nonisomorphic graphs is the graph-theoretic counterpart of medieval knights searching for the "holy grail."[4] For the moment, let us observe that the numbers of vertices and edges, together with the degree sequence, do not comprise such a list.

1.11 Example. The graphs illustrated in Fig. 1.7 both have 5 vertices, 4 edges, and degree sequence (2, 2, 2, 1, 1), yet they cannot be isomorphic: Suppose f were an isomorphism from P_5 to H. If u and v are the vertices of P_5 of degree 1, then $f(u)$ and $f(v)$ would have to be the degree 1 vertices of H. Because $uv \notin E(P_5)$, it would have to be that $f(u)f(v) \notin E(H)$. Since the vertices of degree 1 of H are, in fact, adjacent, f cannot exist. □

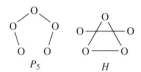

Figure 1.7

Let us call a list of invariants *complete* if it suffices to distinguish nonisomorphic graphs, and *independent* if no invariant on the list can be determined from the others. Example 1.11 shows that a list comprised of the number of vertices, the number of edges, and the degree sequence is not complete. Neither is it independent. The number of vertices is equal to the number of entries in the sequence $d(G)$, and the number of edges can be determined from $d(G)$ by what has come to be known as the "first theorem of graph theory."

1.12 Theorem. *Let $G = (V, E)$ be a graph with vertex set $V = \{v_1, v_2, \ldots, v_n\}$ and edge set E of cardinality $o(E) = m$. Then*

$$\sum_{i=1}^{n} d(v_i) = 2m.$$

Proof. By definition, $d(v)$ is the number of edges that are incident with vertex v. Thus, in summing $d(v)$, each edge is counted twice, once at each of its vertices. ∎

Taking nothing away from the proof in Example 1.11, there is a more obvious reason way the graphs in Fig. 1.7 are not isomorphic; that is, P_5 is "connected"

---

[4] Discussions of computational intractability frequently involve the class NP of decision problems that are solvable "nondeterministically" in polynomial time. In 1971, S. A. Cook proved that every problem in NP can be reduced (in polynomial time) to the "satisfiability" problem, making it the first problem proven to be *NP-complete*. In 1972, R. M. Karp asked whether the Graph Isomorphism Problem is NP-complete. As of this writing, the answer to Karp's question is still not known.

and H is not. Unfortunately, like many "obvious" notation, connectedness is easier to see than it is to define.

1.13 Definition. Let $G = (V, E)$ be a graph, and suppose $u, w \in V$. A *path in* G of length r, from u to w, is a sequence $[v_0, v_1, \ldots, v_r]$ of distinct vertices such that $u = v_0$, $w = v_r$, and $\{v_{i-1}, v_i\} \in E$, $1 \leq i \leq r$.

If, in Definition 1.13, $V(G) = \{v_0, v_1, \ldots, v_r\}$ and $E(G) = \{\{v_i - 1, v_i\}: 1 \leq i \leq r\}$, then $G \cong P_{r+1}$; that is, G, itself, is a path. If G is the Petersen graph in Fig. 1.6, then $[1, 2, 4, 5]$ is a path of length 3. The expressions $[1, 3, 5, 4, 2, 1]$ and $[1, 2, 8, 9]$ are not paths. In the first case, the vertices are not distinct. In the second, $\{2, 8\} \notin E(G)$. (Note that it does not matter which picture, G_1 or G_2, we use to illustrate this discussion!) If $\{u, v\} \in E(G)$, then $[u, v]$ is a path of length 1 from u to v and $[v, u]$ is a path of length 1 from v to u.

1.14 Definition. Let $G = (V, E)$ and $H = (W, F)$ be graphs. If $W \subset V$ and $F \subset E$, then H is a *subgraph* of G. The subgraph of G *induced* by W is $G[W] = (W, E \cap W^{(2)})$.

If $H = (W, F)$ is a subgraph of $G = (V, E)$ then, because H is a graph, $F \subset W^{(2)}$. Thus, $F \subset E \cap W^{(2)}$, with equality if and only if $H = G[W]$. In particular, H is a subgraph of G if and only if H is a subgraph of $G[W]$; in other words, $G[W]$ is the unique maximal subgraph of G having vertex set W. While P_5 is isomorphic to a subgraph of C_5, it is not an induced subgraph; P_4, on the other hand, is isomorphic to an induced subgraph both of P_5 and of C_5.

1.15 Definition. The graph $G = (V, E)$ is *connected* if $o(V) = 1$ or if every pair of distinct vertices of G is joined by a path in G. A (connected) *component* of G is an induced subgraph $H = G[W]$ such that H is connected, but $G[X]$ is not connected for any subset X of V that strictly contains W.

Another way to get at the notion of a component of G is to define a relation among its vertices. Write $u \sim w$ if there is a path in G from u to w. Then "$\sim$" is an equivalence relation on $V(G)$. The induced subgraphs, $G[W]$, as W runs over the equivalence classes of $V(G)$, are the components of G.

Evidently, components are maximal connected subgraphs, and G is connected if and only if it has exactly one component; a graph with more than one component is said to be *disconnected*.

1.16 Theorem. *The number of components is an invariant.*

Proof. Suppose $G = (V, E)$ and $H = (W, F)$ are isomorphic graphs. Let $f: V \to W$ be an isomorphism. If $[v_0, v_1, \ldots, v_r]$ is a path in G from $u = v_0$ to $w = v_r$, then $[f(v_0), f(v_1), \ldots, f(v_r)]$ is a path in H from $f(u)$ to $f(w)$. Thus, u and w are in the same component of G if and only if $f(u)$ and $f(w)$ are in the same component of H. ∎

1.17 Example. The graph H in Fig. 1.7 has two components, one isomorphic to P_2 and the other to C_3. (Don't be misled by the intersection of arcs representing nonadjacent edges of H. An alternate illustration of H appears in Fig. 1.8.) Because H has two components, it cannot be isomorphic to P_5. □

Figure 1.8. Alternate illustration of H.

Suppose $V = \{v_1, v_2, \ldots, v_n\}$ and $W = \{w_1, w_2, \ldots, w_n\}$ are fixed but arbitrary n-element sets. Denote by $\mathcal{V}$ the set of all $2^{C(n,2)}$ (different) graphs with vertex set V, and denote by $\mathcal{W}$ the graphs with vertex set W. Define a function $\varphi: \mathcal{V} \to \mathcal{W}$ as follows: If $G = (V, E) \in \mathcal{V}$, then $\varphi(G) = (W, F)$, where $F = \{w_i w_j : v_i v_j \in E\}$. Then φ establishes a one-to-one correspondence between $\mathcal{V}$ and $\mathcal{W}$ in which corresponding graphs are isomorphic. In the mathematics of graph theory, the nature of the vertices in immaterial. It doesn't matter whether they are cities on a route map, atoms in a chemical molecule, or microprocessors in a computer. In talking about "the graphs on n vertices," it doesn't matter which n vertices, just that there are n of them.

Suppose $o(V) = n$. Denote by $K_n = (V, V^{(2)})$ the graph whose edge set consists of all possible two-element subsets of V. As in Example 1.3, K_n is called the *complete* graph on n vertices. Two illustrations of K_4 can be found in Fig. 1.4. The "complement" of K_n is the graph $K_n^c = (V, \phi)$ having n components, each of which consists of a single *isolated* vertex.

1.18 Definition. Let $G = (V, E)$ be a graph. Its *complement* is the graph $G^c = (V, V^{(2)} \setminus E)$, that is, G and G^c share the same set of vertices, but $uv \in E(G)$ if and only if $uv \notin E(G^c)$.

For $G = (V, E)$ and $H = (W, F)$ to be isomorphic, there must exist an isomorphism $f: V \to W$ such that $uv \in E$ if and only if $f(u)f(v) \in F$. But, that means $uv \notin E$ if and only if $f(u)f(v) \notin F$. Evidently, every isomorphism from G to H is simultaneously an isomorphism from G^c to H^c. In particular, G and H are isomorphic if and only if G^c and H^c are isomorphic.

The nonisomorphic graphs on $n = 4$ vertices are illustrated in Fig. 1.9. With the exception of P_4, these graphs appear in complementary pairs; the path is exceptional because it is isomorphic to its complement, in other words, P_4 is *self-complementary*. The path is also exceptional in the sense that $G = P_4$ is the only graph on 4 vertices with the property that G and G^c are both connected. As we now see, this is not just an accident due to the fact that $n = 4$ is small.

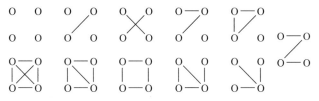

Figure 1.9

1.19 Theorem. *Let G be a graph on $n \geq 2$ vertices. If both G and G^c are connected, then G contains an induced subgraph isomorphic to P_4.*

Note that, because $P_4^c \cong P_4$, G contains an induced subgraph isomorphic to P_4 if and only if G^c contains an induced subgraph isomorphic to P_4. Thus, the theorem is really a statement about connected complementary *pairs* of graphs (cc-pairs for short).

Proof. If the theorem fails, then, from among all counterexamples, there is a "minimal" cc-pair—that is, one having a fewest number, k, of vertices. Because one of G and G^c is disconnected for all graphs G on $2 \leq n \leq 3$ vertices (check it with paper and pencil), k must be at least 4. From Fig. 1.9, we see that the only cc-pair on 4 vertices is $G = P_4 \cong G^c$, and the theorem is true for this pair. Thus, our minimal counterexample is a pair of connected graphs, G and G^c, on the same set V of $k \geq 5$ vertices, and neither G nor G^c contains an induced subgraph isomorphic to P_4.

For a fixed but arbitrary $u \in V$, let $H = G[V \setminus \{u\}]$ be the subgraph of G induced on the remaining vertices. Informally, $H = G - u$ is the graph obtained from G by deleting vertex u and every edge incident with it. Because any induced subgraph of H is an induced subgraph of G, no induced subgraph of H can be isomorphic to P_4. Since H has $k - 1 < k$ vertices, it cannot be a counterexample. Thus, H and H^c cannot both be connected. Without loss of generality, we may assume that H is disconnected.

Because u is not an isolated vertex of G^c, there is a vertex $y \in V$ such that $uy \in E(G^c)$. Let C be the component of H containing y. Suppose $z \in V(H)$ belongs to a component of H different from C. Let $Q = [y = x_0, x_1, \ldots, x_t = z]$ be a shortest path in G from y to z. Because $y \neq u \neq z$, $uy \notin E(G)$, and Q is not a path in H, there is an integer s such that $1 < s < t$ and $u = x_s$. Moreover, since Q is a shortest path, $\{x_{s-2}, x_s\}$, $\{x_{s-2}, x_{s+1}\}$, and $\{x_{s-1}, x_{s+1}\}$ cannot be edges of G. Therefore, $G[x_{s-2}, x_{s-1}, x_s, x_{s+1}] \cong P_4$, contradicting the assumption that G is a counterexample. ∎

1.20 Example. Consider the graphs G_1, G_2, H_1, and H_2 shown in Fig. 1.10. Observe that a pair of numbered vertices is adjacent in G_1 if and only if the same numbered pair is not adjacent in G_2. This proves, by the numbers, that $G_1^c \cong G_2$. Because G_1 and G_1^c are both connected, it follows from Theorem 1.19 that

Chap. 1 Invariants

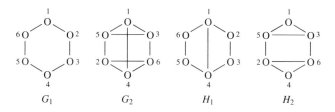

Figure 1.10

G_1 contains an induced subgraph isomorphic to P_4. Indeed, $G_1[1, 2, 3, 4] \cong P_4$. Because G_2 and $G_2^c \cong G_1$ are both connected, G_2 also must have an induced subgraph isomorphic to P_4. Note that $G_2[1, 3, 6, 4] \cong C_4$, not P_4. On the other hand, $P_4 \cong P_4^c \cong (G_1[1, 2, 3, 4])^c \cong G_1^c[1, 2, 3, 4] \cong G_2[1, 2, 3, 4] = G_2[3, 1, 4, 2]$.

Similarly, by the numbers, $H_1^c \cong H_2$. Because H_1 and H_2 are both connected, they must both contain induced subgraphs isomorphic to P_4. Observe that $H_1[2, 1, 4, 5] \cong P_4 \cong H_2[2, 1, 4, 5]$. □

Like most information, graphs can be encoded by means of arrays of 0's and 1's.

1.21 Definition. Let $G = (V, E)$ be a graph with vertex set $V = \{v_1, v_2, \ldots, v_n\}$. The corresponding $n \times n$ *adjacency matrix* $A(G) = (a_{ij})$ is defined by $a_{ij} = 1$ if $v_i v_j \in E$, and $a_{ij} = 0$, otherwise.

It is clear from the definition that $A(G)$ is a symmetric $(0, 1)$-matrix whose main diagonal consists entirely of zeros and whose ith row contains exactly $d_G(v_i)$ ones. Also clear is the fact that $A(G)$ depends not only on G but on the numbering of its vertices.

1.22 Example. If G_1 and G_2 are the graphs illustrated in Fig. 1.11, then

$$A(G_1) = \begin{pmatrix} 0 & 1 & 1 & 1 \\ 1 & 0 & 0 & 0 \\ 1 & 0 & 0 & 1 \\ 1 & 0 & 1 & 0 \end{pmatrix} \quad \text{and} \quad A(G_2) = \begin{pmatrix} 0 & 0 & 0 & 1 \\ 0 & 0 & 1 & 1 \\ 0 & 1 & 0 & 1 \\ 1 & 1 & 1 & 0 \end{pmatrix}.$$

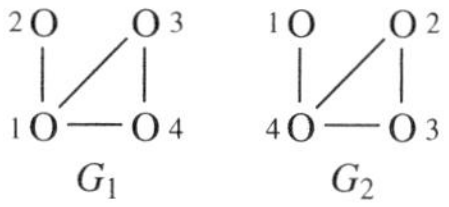

Figure 1.11

Let $f: V(G_1) \to V(G_2)$ be the isomorphism defined by $f(1) = 4$, $f(2) = 1$, $f(3) = 2$, and $f(4) = 3$. The corresponding permutation matrix[5] is defined by $P(f) = (\delta_{if(j)})$, where $\delta_{ik} = 1$ if $i = k$, and 0 otherwise, is the so-called *Kronecker delta*. Note that

$$P(f) = \begin{pmatrix} 0 & 1 & 0 & 0 \\ 0 & 0 & 1 & 0 \\ 0 & 0 & 0 & 1 \\ 1 & 0 & 0 & 0 \end{pmatrix}$$

is the matrix obtained by permuting the rows of the identity matrix I_4 according to the permutation f. The connection between $A(G_1)$ and $A(G_2)$ is given by

$$A(G_1) = P(f)^{-1} A(G_2) P(f),$$

which can be proved by showing that $P(f) A(G_1) = A(G_2) P(f)$. □

In fact, as we now see, two graphs are isomorphic if and only if their adjacency matrices are permutation-similar.

1.23 Theorem. *Graphs G_1 and G_2 are isomorphic if and only if there is a permutation matrix P such that*

$$A(G_1) = P^{-1} A(G_2) P. \tag{1}$$

Proof. If G_1 and G_2 have different numbers of vertices, they cannot be isomorphic and, being of different sizes, their adjacency matrices cannot be similar, much less permutation-similar. Thus, we may assume that G_1 and G_2 have the same number n of vertices. Indeed, without loss of generality we may assume that $V(G_1) = V(G_2) = V = \{1, 2, \ldots, n\}$.

Let $A(G_1) = (a_{ij})$ and $A(G_2) = (b_{ij})$. If $G_1 \cong G_2$, there is a permutation[6] f of V such that $\{i, j\} \in E(G_1)$ if and only if $\{f(i), f(j)\} \in E(G_2)$ — that is, such that $a_{ij} = b_{f(i)f(j)}$, $1 \leq i, j \leq n$. Let $P = P(f) = (\delta_{if(j)})$. Then (Exercise 29) the (i, j)-entry of P^{-1} is $\delta_{jf(i)}$, the (j, i)-entry of P. From the definition of matrix multiplication, the (i, j)-entry of $P^{-1} A(G_2) P$ is

$$\sum_{s,t=1}^{n} (P^{-1})_{is} A(G_2)_{st} P_{tj} = \sum_{s,t=1}^{n} \delta_{sf(i)} b_{st} \delta_{tf(j)}$$

$$= b_{f(i)f(j)}$$

$$= a_{ij}.$$

Therefore, $P^{-1} A(G_2) P = A(G_1)$.

---

[5] An $n \times n$ (0, 1)-matrix is a permutation matrix if it has exactly one 1 in each row and column — that is, if it contains exactly n ones, no two of which lie in the same row or column.

[6] A *permutation* of V is a one-to-one function from V onto V.

Conversely, let P be a permutation matrix that satisfies Equation (1). Define $f(j)$ to be the row index of the unique 1 in column j of P, $1 \le j \le n$. Because P is a permutation matrix, $f: V \to V$ is one-to-one. Indeed, $P = P(f)$. From the definition of $A(G_1)$, $\{i, j\} \in E(G_1)$ if and only if $a_{ij} = 1$. From Equation (1), $a_{ij} = 1$ if and only if $1 = (P^{-1}A(G_2)P)_{ij} = b_{f(i)f(j)}$, if and only if $\{f(i), f(j)\} \in E(G_2)$, proving that f is an isomorphism from G_1 to G_2. ∎

It is a theorem in linear algebra that two $n \times n$, real symmetric matrices are similar if and only if they have the same characteristic polynomial. Together with Theorem 1.23, this proves that the adjacency characteristic polynomial is an invariant: If G_1 and G_2 are isomorphic, then

$$\det(xI_n - A(G_1)) = \det(xI_n - A(G_2)).$$

What about the converse? Can two symmetric $(0, 1)$-matrices be similar without being permutation-similar?

1.24 Example. If G and H are the graphs in Fig. 1.12, then $\det(xI_5 - A(G)) = x^5 - 4x^3 = \det(xI_5 - A(H))$, so $A(G)$ and $A(H)$ have the same characteristic polynomial. Therefore, the real symmetric matrices $A(G)$ and $A(H)$ must be similar. Because $G \not\cong H$, $A(G)$ and $A(H)$ cannot be permutation-similar. □

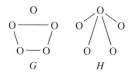

Figure 1.12

Suppose G is a graph on n vertices with two connected components, H_1 and H_2, having n_1 and n_2 vertices, respectively. Then, with a suitable numbering of the vertices of G, $A(G)$ can be expressed as the *direct sum* $A(H_1) \oplus A(H_2)$; that is,

$$A(G) = \begin{pmatrix} A(H_1) & 0 \\ 0 & A(H_2) \end{pmatrix}, \qquad (2)$$

where the upper-right 0 is an $n_1 \times n_2$ submatrix, each of whose entries is zero, and the lower-left block of zeros is $n_2 \times n_1$. It follows from Equation (2) that the characteristic polynomial of $A(G)$ can be factored as a product of the characteristic polynomials of $A(H_1)$ and $A(H_2)$.

This observation is borne out by Example 1.24, where

$$\det(xI_5 - A(G)) = x \times (x^4 - 4x^2)$$
$$= \det(xI_1 - A(K_1)) \times \det(xI_4 - A(C_4)).$$

(On the other hand, despite the fact that the graph H in Example 1.24 is connected, $\det(xI_5 - A(H))$ has the same factorization.)

1.25 Example. Suppose G is a graph on n vertices. Then $A(G^c)$ is the matrix obtained from $A(G)$ by "switching" its *off*-diagonal entries: For all i *different* from j, the (i, j)-entry of $A(G^c)$ is 1 if and only if the (i, j)-entry of $A(G)$ is 0. Alternatively, $A(G^c) = A(K_n) - A(G)$.

If G_1 is the graph from Example 1.22, then G_1^c is a graph with two components, one an isolated vertex (isomorphic to K_1) and the other isomorphic to P_3. With respect to the numbering of the vertices of G_1 illustrated in Fig. 1.11,

$$A(G_1^c) = A(K_4) - A(G_1)$$

$$= \begin{pmatrix} 0 & 1 & 1 & 1 \\ 1 & 0 & 1 & 1 \\ 1 & 1 & 0 & 1 \\ 1 & 1 & 1 & 0 \end{pmatrix} - \begin{pmatrix} 0 & 1 & 1 & 1 \\ 1 & 0 & 0 & 0 \\ 1 & 0 & 0 & 1 \\ 1 & 0 & 1 & 0 \end{pmatrix}$$

$$= \begin{pmatrix} 0 & 0 & 0 & 0 \\ 0 & 0 & 1 & 1 \\ 0 & 1 & 0 & 0 \\ 0 & 1 & 0 & 0 \end{pmatrix}$$

$$= A(K_1) \oplus A(P_3).$$

Similarly,

$$A(G_2^c) = \begin{pmatrix} 0 & 1 & 1 & 0 \\ 1 & 0 & 0 & 0 \\ 1 & 0 & 0 & 0 \\ 0 & 0 & 0 & 0 \end{pmatrix}$$

$$= A(P_3) \oplus A(K_1).$$

Because the permutation $f: \{1, 2, 3, 4\} \to \{1, 2, 3, 4\}$ discussed in Example 1.22 is an isomorphism from G_1 to G_2, it is an isomorphism from G_1^c to G_2^c. In particular,

$$A(G_1^c) = P(f)^{-1} A(G_2^c) P(f),$$

where (as in Example 1.22)

$$p(f) = \begin{pmatrix} 0 & 1 & 0 & 0 \\ 0 & 0 & 1 & 0 \\ 0 & 0 & 0 & 1 \\ 1 & 0 & 0 & 0 \end{pmatrix}.$$

(Before reading on, confirm that $p(f)A(G_1^c) = A(G_2^c)P(f)$.) □

Chap. 1 Invariants

EXERCISES

1. Let $V = \{1, 2, 3, 4\}$. Find a set $E \subset V^{(2)}$ such that $G = (V, E)$ is illustrated by

   (a) ⟨graph⟩ (b) ⟨graph⟩ (c) ⟨graph⟩

   (*Hint*: There are many correct answers.)

2. Among the graphs illustrated in Fig. 1.9, draw the one that is isomorphic to

   (a) P_4 (b) K_4 (c) ⟨graph⟩ (d) ⟨graph⟩ (e) ⟨graph⟩

3. Illustrate a graph with degree sequence (3, 3, 3, 1, 1, 1).

4. Prove the following:

   (a) Every graph is isomorphic to itself.

   (b) If G_1 is isomorphic to G_2, then G_2 is isomorphic to G_1.

   (c) If G_1 is isomorphic to G_2 and G_2 is isomorphic to G_3, then G_1 is isomorphic to G_3.

   (d) "Isomorphic to" is an equivalence relation on graphs.

5. Of the 11 nonisomorphic graphs on 4 vertices, how many are connected?

6. Use Fig. 1.9 to prove that two graphs on four vertices are isomorphic if and only if they have the same degree sequence.

7. Let G be a graph with n vertices and m edges. Prove that $\delta(G) \leq 2m/n \leq \Delta(G)$.

8. Illustrate four (nonisomorphic) graphs,

   (a) each having degree sequence (3, 3, 2, 2, 1, 1).[7]

   (b) each having degree sequence (4, 3, 2, 2, 2, 1).

9. Explain why no graph can have degree sequence

   (a) (4, 3, 2, 2, 1, 1). (b) (6, 4, 3, 3, 2, 2).

10. Illustrate six (nonisomorphic) graphs, each having five vertices and five edges.

---

[7] A theoretical procedure for finding all nonisomorphic graphs with a prescribed degree sequence is outlined in [R. Grund, Konstruktion schlichter Graphen mit gegebener Gradpartition, *Bayreuther Math. Schriften* **44** (1993), 73–104].

11 Illustrate the complement of

(a) (b)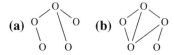

12 Recall that a graph is self-complementary if it is isomorphic to its complement.

(a) Find a self-complementary graph on 5 vertices.

(b) Prove that no graph on 6 vertices is self-complementary.

13 Let G be a graph on n vertices with degree sequence
$d(G) = (d_1, d_2, \ldots, d_n)$. Show that

$$d(G^c) = (n - 1 - d_n, \ldots, n - 1 - d_2, n - 1 - d_1).$$

14 Prove that $(G^c)^c = G$.

15 Let $G = (V, E)$ be a graph having k odd vertices; that is, $k = o(\{v \in V : d(v) \text{ is odd}\})$. Prove that k is even.

16 Recall that a disconnected graph is a graph with more than one component.

(a) Show that the complement of a disconnected graph is connected.

(b) Illustrate a disconnected graph G that has an induced subgraph isomorphic to P_4.

17 Let G be a graph whose vertices and edges are the 8 vertices and 12 edges of a cube, respectively. Prove (by the numbers) that G is isomorphic to the graph illustrated in Fig. 1.13.

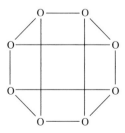

Figure 1.13

18 Prove that the graph illustrated in Fig. 1.14 is isomorphic to the Petersen graph pictured (twice) in Fig. 1.6.

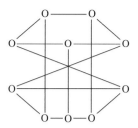

Figure 1.14

19 A graph is *r-regular* (or *regular* of *degree r*) if each of its vertices has degree r. Prove that the 3-regular graphs in Fig. 1.15 are not isomorphic.

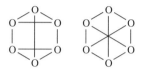

Figure 1.15

20 A graph on n vertices is *antiregular* if its multiset of vertex degrees contains $n - 1$ different numbers.

  (a) Illustrate the (nonisomorphic) connected antiregular graph(s) on 5 vertices.

  (b) Illustrate the (nonisomorphic) connected antiregular graph(s) on 6 vertices.

  (c) Suppose G is a connected antiregular graph on $n \geq 2$ vertices. Prove that

  $$d(G) = (n - 1, n - 2, \ldots, n - r + 1, n - r, n - r,$$
  $$n - r - 1, \ldots, 2, 1),$$

  where $r = \lfloor (n + 1)/2 \rfloor$, the integer part of $(n + 1)/2$.

  (d) Let G be a connected antiregular graph on $n \geq 2$ vertices. Prove that the number of edges of G is

  $$m = \begin{cases} n^2/4 & \text{if } n \text{ is even,} \\ (n^2 - 1)/4 & \text{if } n \text{ is odd.} \end{cases}$$

[*Hint*: Use part (c).]

21 Let G be a connected graph on $n \geq 2$ vertices. Suppose G does not have an induced subgraph isomorphic to P_4 or C_4. Prove that G has a dominating vertex, in other words, that $\Delta(G) = n - 1$.

22 A *multigraph* consists of two things: (1) a nonempty (vertex) set V and (2) a multiset E with the property that every element of E is an element of $V^{(2)}$. So, a multigraph is like a graph except that more than one edge can be incident to the same pair of vertices.

  (a) Draw the multigraph $M = (V, E)$, where $V = \{a, b, c, d\}$ and $E = \{ab, ab, ad, bc, bc, bd, cd\}$.

  (b) Show that Theorem 1.12 remains valid for multigraphs.

  (c) Define "isomorphism" for multigraphs.

23 Let G be the graph H_1 in Fig. 1.10. Find all induced subgraphs of G that are isomorphic to P_4.

24 Let G be a graph with n vertices and m edges. Show that exactly m 1's lie above the main diagonal of $A(G)$.

25 For the graphs in Example 1.22, compute

  (a) $\det(xI_n - A(G_1))$ **(b)** $\det(xI_n - A(G_2))$

26 Confirm that $\det(xI_4 - A(C_4)) = x^4 - 4x^2$.

27 Show that $\det(xI_6 - A(G)) = \det(xI_6 - A(H))$ for the graphs G and H illustrated in Fig. 1.16.

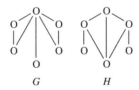

Figure 1.16

28 Let G be a graph having n vertices and m edges. Prove that the coefficient of x^{n-2} in the characteristic polynomial of $A(G)$ is $-m$.

29 Let f be a permutation of $V = \{1, 2, \ldots, n\}$, that is, a one-to-one function from V onto V. Let $P(f)$ be the $n \times n$ matrix whose (i, j)-entry is $\delta_{if(j)}$. Show that

  (a) $P(f)^{-1} = P(f^{-1})$.

  (b) $P(f)^{-1} = P(f)^t$, the *transpose*[8] of $P(f)$.

---

[8] The transpose of $A = (a_{ij})$ is the matrix A^t whose (i, j)-entry is a_{ji}, the (j, i)-entry of A.

30 A *walk of length k* in G, from v_i to v_j, is a sequence of its vertices $v_i = u_0, u_1, u_2, \ldots, u_k = v_j$ such that $u_{i-1}u_i \in E(G)$, $1 \le i \le k$. (A path is a walk in which the u's are all distinct.) Prove that the (i, j)-entry of $A(G)^k$ is the number of walks of length k in G from v_i to v_j.

31 Let S be a finite set. Suppose $\mathscr{F} = \{S_i \subset S: 1 \le i \le n\}$ is a family of subsets of S, where we allow the possibility that $S_i = S_j$ when $i \ne j$. The corresponding *intersection graph* is $G = (\mathscr{F}, \mathscr{E})$, where $S_i S_j \in \mathscr{E}$ if (and only if) $i \ne j$ and $S_i \cap S_j \ne \phi$. If $S = \{a, b, c\}$, then K_2^c is an intersection graph corresponding, for example, to $\mathscr{F} = \{\{a\}, \{b, c\}\}$ or to $\mathscr{F} = \{\phi, \phi\}$; K_2 corresponds to $\mathscr{F} = \{\{a\}, \{a\}\}$ or to $\mathscr{F} = \{\{a, b\}, \{b, c\}\}$, P_3 to $\mathscr{F} = \{\{a\}, \{a, b\}, \{b, c\}\}$, and so on. [9] The *intersection number*, $\mathbf{i}(G)$, is the minimum cardinality of a set S such that G is an intersection graph based on a family of subsets of S. Evidently, $\mathbf{i}(K_2^c) = 0, \mathbf{i}(K_2) = 1, \mathbf{i}(P_3) = 2$, and so on.

(a) Show that $\mathbf{i}(G) = 3$ for the graph G in Fig. 1.16.

(b) Show that $\mathbf{i}(H) = 4$ for the graph H in Fig. 1.16.

32 Let G be a connected graph having $n \ge 3$ vertices and m edges. In the language of Exercise 31, show that

(a) G is an intersection graph. (Hint: Let $S = E(G)$.)

(b) the intersection number, $\mathbf{i}(G) \le m$.

33 A *binary codeword of length k* is a k-letter "word" made from the "alphabet" $\{0, 1\}$. Thus, for example, $\{00, 01, 10, 11\}$ is the set of all binary codewords of length 2. The *Hamming distance* between two binary codewords of length k is the number of places in which they differ. So, the distance between 01 and 10 is 2 while the distance between 01 and 11 is 1. The vertex set of the *cube graph*, Q_k, is the set of all binary codewords of length k. Two codewords are adjacent in Q_k if (and only if) the distance between them is 1.

(a) Show that Q_k has $n = 2^k$ vertices.

(b) Show that $Q_2 \cong C_4$.

(c) Show that Q_3 is illustrated by the vertices and edges of a cube.

(d) Show that Q_k is k-regular. (See Exercise 19.)

(e) Find the number m of edges in Q_k.

34 Denote by $g(n, m)$ the number of nonisomorphic graphs having n vertices and m edges and define the *generating function* $f_n(x) = \sum_{m \ge 0} g(n, m) x^m$.

(a) Show that $f_4(x) = 1 + x + 2x^2 + 3x^3 + 2x^4 + x^5 + x^6$.

---

[9] Terminology varies: Some authors require the subsets comprising $\mathscr{F}$ to be nonempty and distinct.

(b) Show that $f_n(x)$ is a monic polynomial of degree $C(n, 2)$.

(c) If $f_n(x) = c_0 + c_1 x + \cdots + c_k x^k$, show that $c_r = c_{k-r}$, $0 \le r \le C(n, 2)$.

35 Let G and H be the graphs from Example 1.24 (illustrated in Fig. 1.12).

(a) Show that $\det(A(G^c)) \ne 0 = \det(A(H^c))$.

(b) Show that $\det(xI_5 - A(G^c)) \ne \det(xI_5 - A(H^c))$.

2
Chromatic Number

In this chapter we are going to decorate graphs by coloring their vertices. The idea is most easily described using the language of functions: A *coloring* (or *r-coloring*) of G is a function from $V(G)$ into some set C of r colors.

2.1 Definition. A *proper* coloring of G is one in which adjacent vertices are colored differently. Denote the number of proper r-colorings of G by $p(G, r)$.

Let G be a graph and C a set of colors. The evaluation of $p(G, r)$ involves counting functions $f: V(G) \to C$ that satisfy $f(u) \neq f(v)$ for all $uv \in E(G)$. Frequently useful in this sort of process is the following self-evident principle.

2.2 Fundamental Counting Principle. *Consider a sequence of two decisions. If the number of choices for the second does not depend on how the first is made, then the number of ways to make both decisions is the product of the numbers of choices for each of them.*

In symbols, suppose there are s ways to make the first decision and, independently, t ways to make the second. Then the number of ways to make the sequence of two decisions is $s \times t$. What about n decisions? If c_{i+1}, the number of ways to make decision $i + 1$, is independent of how the previous i decisions are made, then the number of ways to make the sequence of n decisions is the product $c_1 \times c_2 \times \cdots \times c_n$.

Let's apply this principle to vertex colorings. Suppose G is a graph with n vertices, and C is a set of r colors. Coloring $V(G)$ corresponds to a sequence of n decisions, each having r choices. Hence, the number of colorings $f: V(G) \to C$ is $r \times r \times \cdots \times r = r^n$. Included in this count are the constant functions, corresponding to monochromatic colorings. Of course, if G has so much as one edge, no such coloring can be proper. On the other hand, if G has no edges, then every coloring is proper; that is,

$$p(K_n^c, r) = r^n. \qquad (3)$$

Because it has n vertices, selecting an r-coloring for the complete graph, K_n, also corresponds to a sequence of n decisions, and there are still r choices for the first of them. If the coloring is going to be proper, however, the number of choices[1] for the second decision is $r-1$. There are $r-2$ for the third decision, and so on. Thus,

$$p(K_n, r) = r^{(n)}$$
$$= r \times (r-1) \times \cdots \times (r-n+1), \qquad (4)$$

the so-called *falling factorial* function. Note, in particular, that $p(K_n, r) = 0$, for all $r < n$. But, never mind the Fundamental Counting Principle, it follows directly from the definition that $p(G, r) = 0$ whenever G contains a set of $t > r$ mutually adjacent vertices.

A *clique* (in G) is a nonempty set of pairwise adjacent vertices. Thus, $\phi \neq W \subset V(G)$ is a clique if and only if $W^{(2)} \subset E(G)$, if and only if the induced subgraph $G[W] \cong K_t$, where $t = \mathrm{o}(W)$. There is a complementary notion: An *independent set* (in G) is a nonempty set of mutually *non*adjacent vertices. So, $\phi \neq W \subset V(G)$ is an independent set if and only if no two of its vertices are adjacent, if and only if $G[W] \cong K_t^c$, if and only if W is a clique in G^c.

2.3 Definition. Let G be a graph. The *clique* number, $\omega(G)$, is the maximum number of vertices in any clique of G. The *independence* number, $\alpha(G) = \omega(G^c)$, is the number of vertices in a largest independent set of G.

2.4 Example. Let $G = C_n$ (the cycle on n vertices). Then $\alpha(G) = \lfloor n/2 \rfloor$, the integer part of $n/2$, and

$$\omega(G) = \begin{cases} 3 & \text{if } n = 3, \\ 2 & \text{if } n \geq 4. \end{cases}$$

(Draw some pictures and confirm these statements for C_3, C_4, C_5, and C_6.) □

As we have observed, $p(G, r) = 0$ whenever $r < \omega(G)$. So, suppose $r \geq \omega(G)$ and assume $f: V(G) \to C$ is a proper r-coloring. For each $c \in C$, define $f^{-1}(c) = \{v \in V(G) : f(v) = c\}$. The *color class*, $f^{-1}(c)$, consists of those vertices of G to which f assigns the color c. From the definition of proper coloring, $f^{-1}(c)$ is an independent set. Therefore, $\mathrm{o}(f^{-1}(c)) \leq \alpha(G)$, $c \in C$. No single color can be shared by more than $\alpha(G)$ vertices. Because $V(G)$ is the disjoint union of the color classes,

$$n = \mathrm{o}(V(G))$$
$$= \sum_{c \in C} \mathrm{o}(f^{-1}(c))$$
$$\leq r\alpha(G).$$

---
[1] While the choices themselves may depend on how the first decision is made, their *number* does not, and this is all that matters in applications of the Fundamental Counting Principle.

2.5 Definition. Let G be a graph. Its *chromatic number*, $\chi(G)$, is the smallest number of colors that suffice to color G properly.[2]

Evidently, $p(G, r) = 0$ for all $r < \chi(G)$ and $p(G, r) > 0$ for all $r \geq \chi(G)$. In terms of $\chi(G)$, our most recent observations may be summarized as follows:

2.6 Theorem. *If G is a graph on n vertices, then*
(a) $\chi(G) \geq \omega(G)$ *and*
(b) $n \leq \chi(G)\alpha(G)$.

Because $\chi(G)$ is an integer, Theorem 2.6b implies that $\chi(G) \geq \lceil n/\alpha(G) \rceil$, the smallest integer not less than $n/\alpha(G)$ — that is, $n/\alpha(G)$ *rounded up*.

2.7 Example. Let $G = C_n$. Because no graph with an edge can be colored properly with one color, $\chi(G) \geq 2$. When n is even, a proper 2-coloring of G can be achieved by alternating the two colors around the cycle. When n is odd, G can be colored properly with three colors, but not with two. Thus,

$$\chi(C_n) = \begin{cases} 2 & \text{if } n \text{ is even,} \\ 3 & \text{if } n \text{ is odd,} \end{cases}$$
$$= \lceil n/\alpha(G) \rceil. \qquad \square$$

2.8 Definition. Suppose $G = (V, E)$ and $H = (W, F)$ are graphs on disjoint sets of vertices. The *union* of G and H is $G \oplus H = (V \cup W, E \cup F)$. If $H \cong G$, then $G \oplus H$ may be written as $2G$. The *join*, $G \vee H$, is the graph obtained from $G \oplus H$ by adding edges from each vertex of G to every vertex of H; that is, $V(G \vee H) = V \cup W$ and $E(G \vee H) = E \cup F \cup \{vw : v \in V \text{ and } w \in W\}$.

If G and H are connected, then $G \oplus H$ is a disconnected graph with two components, one isomorphic to G and the other to H. If $G = K_2$, then $G \vee G \cong K_4$ and $G^c \vee G^c \cong C_4 \cong (2G)^c$. Indeed, unions, joins, and complements are related by the formula

$$G \vee H = (G^c \oplus H^c)^c. \qquad (5)$$

While verifying the following relations has been left to the exercises, why not confirm some of them now?

$$\omega(G \oplus H) = \max\{\omega(G), \omega(H)\}, \qquad (6a)$$
$$\omega(G \vee H) = \omega(G) + \omega(H), \qquad (6b)$$
$$\alpha(G \oplus H) = \alpha(G) + \alpha(H), \qquad (7a)$$
$$\alpha(G \vee H) = \max\{\alpha(G), \alpha(H)\}, \qquad (7b)$$

---

[2] Computing $\chi(G)$ is one of the classic NP-complete problems.

and

$$\chi(G \oplus H) = \max\{\chi(G), \chi(H)\}, \tag{8a}$$

$$\chi(G \vee H) = \chi(G) + \chi(H). \tag{8b}$$

If all n vertices of G are given different colors, the result will be a proper coloring. Hence, $n \geq \chi(G)$. The next notion is useful in obtaining an improvement of this trivial upper bound.

2.9 Definition. Suppose G is a graph. Let $v \in V(G)$. The set of *neighbors* of v is $N_G(v) = \{u \in V(G) : uv \in E(G)\}$.

Evidently, $o(N_G(v)) = d_G(v)$, the degree of $v \in V(G)$. Implicit in Definition 2.9 is that two vertices of G are neighbors, if and only if they are adjacent.

2.10 Theorem. *The chromatic number* $\chi(G) \leq \Delta(G) + 1$.

Note that equality holds when G is a complete graph or an odd cycle.[3]

Proof. Suppose we are given $r = \Delta(G) + 1$ colors. Let H be a fixed but arbitrary component of G and suppose $v \in V(H)$. Because $d(v) = d_H(v) = d_G(v) \leq \Delta(G)$, it is possible to assign different colors to the $d(v) + 1$ vertices in $\{v\} \cup N_G(v)$. Denote by $U \subset V(H)$ the set of vertices that have been colored so far. If $U = V(H)$, we are finished. Otherwise, because H is connected, there is an edge $\{u, w\} \in E(H)$ such that $u \in U$ and $w \notin U$. Because $d(w) \leq \Delta(G)$, there is a color available for w that is not in use by any of its (already colored) neighbors. Replace U with $U \cup \{w\}$ and continue this "greedy" algorithm until $U = V(H)$, thus proving that $\chi(H) \leq \Delta(G) + 1$. Because H was an arbitrary component of G, the result follows from the obvious generalization of Equation (8a) to the union of k graphs. ∎

Apart from the qualitative statement that $p(G, r) > 0$ for all $r \geq \chi(G)$, we have not made much progress on the problem of evaluating $p(G, r)$. Suppose H_1 and H_2 are graphs on disjoint sets of n_1 and n_2 vertices, respectively. Then we may view the proper r-colorings of $G = H_1 \oplus H_2$ from the perspective of $n = n_1 + n_2$ decisions. Or, we may view them in terms of two decisions, the first involving r-colorings of H_1 and the second involving r-colorings of H_2. Because the number of proper colorings of H_2 does not depend on how H_1 might be colored, we have, from the Fundamental Counting Principle, that

$$p(H_1 \oplus H_2, r) = p(H_1, r)p(H_2, r), \tag{9}$$

---

[3] An *odd (even)* cycle is one with an odd (even) number of vertices. It was proved by R. L. Brooks [On coloring the nodes of a network, *Proc. Cambridge Philos. Soc.* **37** (1941), 194–197] that $\chi(G) \leq \Delta(G)$ unless G is a complete graph or an odd cycle.

Chap. 2 Chromatic Number

with the obvious generalization to the union of k graphs. There is another, less obvious generalization of Equation (9). It involves the situation in which there is some "overlap" between H_1 and H_2.

2.11 Definition. Let H_1 and H_2 be graphs on disjoint sets of vertices. Suppose $U = \{u_1, u_2, \ldots, u_t\}$ is a clique in H_1 and $W = \{w_1, w_2, \ldots, w_t\}$ is a clique in H_2. Let G be the graph obtained from $H_1 \oplus H_2$ by identifying (*coalescing* into a single vertex), u_i and w_i, $1 \leq i \leq t$. Then G is an *overlap* of H_1 and H_2 in K_t.

2.12 Example. Graphs G_1 and G_2 in Fig. 2.1 are nonisomorphic overlaps of H_1 and H_2 in K_4; G_1 and G_2 may also be viewed as overlaps of H_2 and K_3 in K_2. □

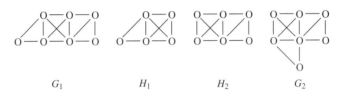

Figure 2.1

2.13 Theorem. *If G is an overlap of H_1 and H_2 in K_t, then $p(G, r) = p(H_1, r) \cdot p(H_2, r)/r^{(t)}$.*

Proof. Suppose $U = \{u_1, u_2, \ldots, u_t\}$ is a clique in H_1. Properly coloring H_1 may be viewed as a sequence of two decisions. The first, how to color the vertices of U, involves $r^{(t)}$ choices. (See Equation (4).) The second decision is how to color the remaining vertices of H_1. Because the vertices of U must all be colored differently, the number of choices for the second decision, call it $q(r)$, is independent of how the first decision is made; that is, $p(H_1, r) = r^{(t)} \times q(r)$. No matter how the first decision is made, $q(r) = p(H_1, r)/r^{(t)}$ is the (integer) number of ways to make the second decision.

Properly coloring G can now be viewed as a sequence of three decisions. The overlapping vertices may be colored in $r^{(t)}$ ways, the remaining vertices of H_1 in $p(H_1, r)/r^{(t)}$ ways, and, independently of how the first two decisions are made, the remaining vertices of H_2 may be colored in $p(H_2, r)/r^{(t)}$ ways. So, by the Fundamental Counting Principle,

$$p(G, r) = r^{(t)} \times \frac{p(H_1, r)}{r^{(t)}} \times \frac{p(H_2, r)}{r^{(t)}}. \qquad \blacksquare$$

2.14 Example. Because

$$G = \begin{array}{c} \circ\!\!-\!\!\circ \\ |\!\!\diagup\!\!| \\ \circ\!\!-\!\!\circ \end{array}$$

is the overlap of K_3 with itself in K_2, we have, from Theorem 2.13, that $p(G, r) = r^{(3)}r^{(3)}/r^{(2)} = r(r-1)(r-2)^2$. Evidently, $\chi(G) = 3$. The $p(G, 3) = 6$ different red–white–blue proper colorings of G are illustrated in Fig. 2.2.

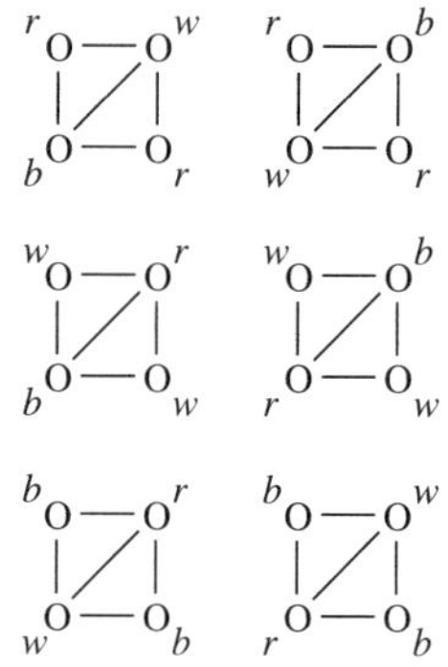

Figure 2.2

Observe that the graphs

$$G_1 = \quad \text{and} \quad G_2 =$$

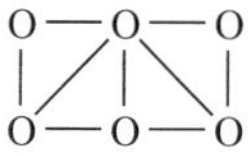

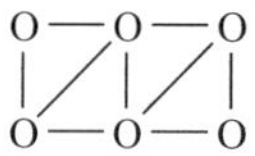

are nonisomorphic overlaps of G with itself in K_2. Therefore, $p(G_1, r) = p(G, r)^2/r^{(2)} = r(r-1)(r-2)^4 = p(G_2, r)$. Graphs like G_1 and G_2 that share the same chromatic polynomial are said to be *chromatically equivalent*. □

Evaluating $p(G, r)$ for general graphs involves a new idea.

2.15 Definition. Let $G = (V, E)$ be a graph and suppose $e = uv \in E$. The *edge subgraph* $G - e = (V, E \setminus \{e\})$ is obtained from G by deleting edge e.

If e is any one of the six edges of K_4, then $G = K_4 - e$ appeared in Example 2.14. If $e = uv$ is the edge of G illustrated in Fig. 2.3(a), then $G - e$ is shown in Fig. 2.3(b).

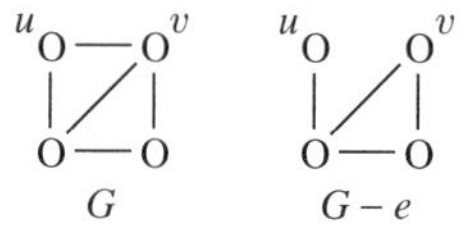

G
(a)

$G - e$
(b)

Figure 2.3

Because vertices u and v remain when edge $e = uv$ is deleted, $G - e$ can never be an induced subgraph of G.

Observe that every proper coloring of G is a proper coloring of $G - e$. The difference, $p(G - e, r) - p(G, r)$, is the number of proper colorings of $G - e$ in which u and v are colored the same. To evaluate this difference, consider the "multigraph" $G|e$ obtained from $G - e$ by identifying u and v (i.e., coalescing u and v into a single vertex). For the situation illustrated in Fig. 2.3, $G|e$ is shown in Fig. 2.4(a). Because there is a one-to-one correspondence between proper colorings of $G|e$ and those proper colorings of $G - e$ in which u and v are colored the same, $p(G - e, r) - p(G, r) = p(G|e, r)$.

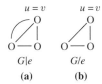

Figure 2.4

Since multiple edges are irrelevant in the enumeration of proper colorings, we have $p(G|e, r) = p(G/e, r)$, where G/e is the *contraction* pictured in Fig. 2.4(b). Hence, $p(G - e, r) - p(G, r) = p(G/e, r)$. Rearranging terms in this equation produces the following fundamental result.

2.16 Chromatic Reduction Theorem. *Let G be a graph. If $e = uv \in E(G)$, then*

$$p(G, r) = p(G - e, r) - p(G/e, r), \qquad (10)$$

where the contraction, G/e, is the graph obtained from $G - e$ by identifying vertices u and v and deleting any redundant edges that may arise in the process.

2.17 Example. Let $G = K_4 - e$. From Example 2.14,

$$p(G, r) = r(r - 1)(r - 2)^2$$
$$= r^4 - 5r^3 + 8r^2 - 4r. \qquad (11)$$

Let's confirm this result using chromatic reduction. Equation (10) can be expressed in the picturesque form where $p(G, r)$ is represented by G. Further applications of Equation (10) produce

$$\begin{pmatrix}\text{□}\end{pmatrix} = \begin{pmatrix}\text{△}\end{pmatrix} - 2\begin{pmatrix}\text{△ with one edge}\end{pmatrix} \quad (12)$$

$$= \left(\begin{pmatrix}\text{path}_3\end{pmatrix} - \begin{pmatrix}\text{edge + isolated}\end{pmatrix}\right) - 2\left(\begin{pmatrix}\text{edge + isolated}\end{pmatrix} - \begin{pmatrix}\text{two isolated}\end{pmatrix}\right)$$

$$= \left(\begin{pmatrix}\text{3 iso}\end{pmatrix} - 2\begin{pmatrix}\text{edge+iso}\end{pmatrix}\right) - 2\left(\begin{pmatrix}\text{edge+iso}\end{pmatrix} - 2\begin{pmatrix}\text{iso}\end{pmatrix}\right).$$

After consolidating isomorphic graphs, the last equation becomes

$$\begin{pmatrix}\text{□}\end{pmatrix} = \begin{pmatrix}\text{3 iso}\end{pmatrix} - 4\begin{pmatrix}\text{edge+iso}\end{pmatrix} + 4\begin{pmatrix}\text{iso}\end{pmatrix}.$$

Another step, consolidation included, produces

$$G = K_4^c - 5K_3^c + 8K_2^c - 4K_1.$$

Finally, because $p(K_n^c, r) = r^n$, we obtain

$$p(G, r) = r^4 - 5r^3 + 8r^2 - 4r, \quad (13)$$

confirming Equation (11).

It is, of course, not generally necessary to carry chromatic reduction all the way to the point at which no edges remain. Applying Equations (4) and (9) to Equation (12), for example, produces $p(G, r) = r \times r^{(3)} - 2r^{(3)} = (r - 2)r^{(3)} = r(r - 1)(r - 2)^2$. □

Suppose G is any graph with n vertices and m edges. If $e \in E(G)$, then $G - e$ has n vertices and $m - 1$ edges, while G/e has $n - 1$ vertices and at most $m - 1$ edges. After sufficiently many applications of chromatic reduction, every edge can be deleted, resulting in an expression of the form

$$p(G, r) = p(K_n^c, r) - c_1 p(K_{n-1}^c, r) + c_2 p(K_{n-2}^c, r) - \cdots$$
$$= r^n - c_1 r^{n-1} + c_2 r^{n-2} - \cdots,$$

a polynomial in r. It seems we have proved the following.

2.18 Corollary. *If G is a graph on n vertices, then $p(G, r)$ is a monic polynomial in r, of degree n, with integer coefficients.*

Chap. 2   Chromatic Number

Recall that $G_1 \cong G_2$ only if $A(G_1)$ and $A(G_2)$ share the same characteristic polynomial. Setting out to enumerate proper colorings, we have stumbled onto a second polynomial invariant. Might a complete list of invariants be comprised of polynomials? (A precedent exists in linear algebra where the so-called similarity invariants are polynomials.) In any case, now that we know $p(G, r)$ is a polynomial, we may as well replace r with the more customary variable x.

2.19 Definition.   The polynomial $p(G, x)$ is called the *chromatic polynomial* of G.

Let G be a graph with chromatic polynomial $f(x) = p(G, x)$. If $k = \chi(G) - 1$, then $f(0) = f(1) = \cdots = f(k) = 0$; that is $(x - i)$ is a factor of $f(x)$, $0 \le i \le k$. Hence, there exist positive integer exponents, $e_0, e_1, \ldots, e_k$, and a polynomial $q(x)$, with integer coefficients, such that

$$f(x) = x^{e_0}(x-1)^{e_1}(x-2)^{e_2}\ldots(x-k)^{e_k}q(x), \tag{14}$$

where $q(x)$ has no integer roots in the interval $[0, k]$. In fact, because G can be colored properly with $\chi(G) = k + 1$ colors, it can be colored properly with $k + i$ colors for all $i \ge 1$. Thus, $f(s) > 0$ for every integer $s > k$, meaning that $q(x)$ cannot have *any* nonnegative integer zeros. If

$$G = \begin{array}{c} \circ\!\!-\!\!\circ \\ |\!\diagup\!| \\ \circ\!\!-\!\!\circ \end{array},$$

for example, then (Equation (11))

$$\begin{aligned} f(x) &= x^4 - 5x^3 + 8x^2 - 4x \\ &= x(x-1)(x-2)^2, \end{aligned}$$

and $q(x) = 1$.

Every graph has at least one vertex, so no graph can be colored (properly or not) with zero colors. The only graphs that can be colored properly with just one color are the complements of complete graphs — that is, collections of isolated vertices. For any graph G with an edge, $\chi(G) \ge 2$.

2.20 Definition.   Graph G is *bipartite*[4] if $\chi(G) \le 2$.

Suppose $G = (V, E)$ is a bipartite graph on n vertices. If $\chi(G) = 1$, then $G = K_n^c$, and there is little more to be said. If $\chi(G) = 2$, then there is at least one proper coloring $f: V(G) \to \{c_1, c_2\}$. Consider the color classes $V_i = f^{-1}(c_i)$, $i = 1, 2$. Then, $V_1 \ne \phi \ne V_2$, $V = V_1 \cup V_2$, and $V_1 \cap V_2 = \phi$. Moreover, V_1 and

---

[4] In the chemical literature, bipartite graphs correspond to the so-called *alternant* hydrocarbons.

V_2 are independent sets. Conversely, if $V(G)$ can be partitioned into two independent sets V_1 and V_2, then $f: V(G) \to \{c_1, c_2\}$ defined by $f(v) = c_i$, $v \in V_i$, $i = 1, 2$, is a proper 2-coloring of G. It seems that, apart from $G = K_1$, a graph is bipartite if and only if its vertex set can be *partitioned* into the disjoint union of two nonempty independent sets. In the combinatorial literature, subsets of a partition are called its *parts* — hence the name "bipartite," meaning two parts. (There may be more than one way to *bipartition* the vertex set of a bipartite graph.)

Let $G = (V, E)$ be a bipartite graph with bipartition $V = V_1 \cup V_2$. Suppose $V_1 = \{v_i : 1 \le i \le k\}$ and $V_2 = \{v_{k+j} : 1 \le j \le n-k\}$, where $n = o(V)$. Let $B = B(G)$ be the $k \times (n-k)$ matrix whose (i, j)-entry is 1 if $\{v_i, v_{k+j}\} \in E$, and 0 otherwise. Then the adjacency matrix of G can be expressed (in "partitioned" form) as

$$A(G) = \begin{pmatrix} 0 & B \\ B^t & 0 \end{pmatrix}, \tag{15}$$

where each "0" represents a square submatrix consisting entirely of zeros and B^t is the transpose of B; that is, B^t is the $(n-k) \times k$ matrix whose (i, j)-entry is the (j, i)-entry of B. Conversely, if some numbering of the vertices of G produces an adjacency matrix of the form given in Equation (15), then G is bipartite.

2.21 Definition. Let s and t be positive integers. The *complete* bipartite graph $K_{s,t} = K_s^c \vee K_t^c$.

Suppose X and Y are disjoint sets of cardinalities $o(X) = s$ and $o(Y) = t$. Let $V = X \cup Y$ and $E = \{xy : x \in X \text{ and } y \in Y\}$. Then $G = (V, E) \cong K_{s,t}$. Indeed, the star $S_n \cong K_{1, n-1}$.

2.22 Definition. Suppose $G = (V, E)$ is a graph. Let $k \ge 3$ be an integer. A *cycle* in G of *length* k is a sequence $< v_1, v_2, \ldots, v_k >$ of distinct vertices such that $\{e_i : 1 \le i \le k\} \subset E$, where $e_i = \{v_i, v_{i+1}\}$, $1 \le i < k$, and $e_k = \{v_k, v_1\}$.

If $W = \{v_i : 1 \le i \le k\}$ and $F = \{e_i : 1 \le i \le k\}$, then $H = (W, F)$ is a subgraph of G isomorphic to C_k. If $< v_1, v_2, \ldots, v_k >$ is a cycle in G of length k, then $[v_1, v_2, \ldots, v_k]$ is a path in G of length $k - 1$.

2.23 Theorem. *Let G be a graph on n vertices. Then G is bipartite if and only if it contains no cycles of odd length.*

The proof of Theorem 2.23 requires another new idea.

2.24 Definition. Let u and w be vertices of a connected graph G. The *distance*, $d(u, w)$, is the length of a shortest path in G from u to w.

Proof of Theorem 2.23. Because three colors are required to properly color the vertices of an odd cycle, no bipartite graph can contain one.

Conversely, suppose G contains no odd cycles. If $n \leq 2$, G is bipartite and we are finished. Otherwise, by Equation (8a), it suffices to assume that G is connected. Let $u \in V(G)$ be fixed but arbitrary. Suppose $w \in V(G)$. If $w = u$ or $d(u, w)$ is even, color vertex w blue; if $d(u, w)$ is odd, color it green. This fails to produce a proper 2-coloring of G only if there exist adjacent vertices w_1 and w_2 that are colored the same. Let $P = [x_0, x_1, \ldots, x_r]$ be a shortest path in G from $u = x_0$ to $x_r = w_1$, and let $Q = [y_0, y_1, \ldots, y_s]$ be a shortest path from $u = y_0$ to $w_2 = y_s$. If $\{x_i : 1 \leq i \leq r\} \cap \{y_j : 1 \leq j \leq s\} \neq \phi$, let k be the largest index such that $x_k \in \{y_j : 1 \leq j \leq s\}$. Because P and Q are shortest paths, it must be that $x_k = y_k$ and $\{x_i : k < i \leq r\} \cap \{y_j : k < j \leq s\} = \phi$. If P and Q overlap only at u, set $k = 0$. In either case,

$$C = <x_k, x_{k+1}, \ldots, x_r, y_s, y_{s-1}, \ldots, y_{k+1}>$$

is a cycle in G of length $\ell = (r - k) + 1 + (s - k - 1) + 1 = r + s - 2k + 1$. Because $w_1 = x_r$ and $w_2 = y_s$ are colored the same, $r = d(u, w_1)$ and $s = d(u, w_2)$ have the same parity. Therefore, ℓ is odd, contradicting that G contains no odd cycles. ∎

2.25 Definition. A graph without any cycles is called a *forest*. A connected forest is a *tree*.

Evidently, a forest is a graph each of whose components is a tree.

Several things seem evident just by glancing at the trees in Fig. 2.5. For one thing, the graphs in the figure are increasingly "condensed" when viewed from left to right and top to bottom.

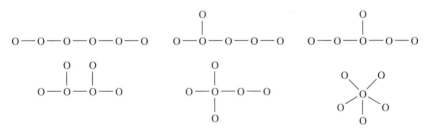

Figure 2.5. The nonisomorphic trees on 6 vertices.

2.26 Definition. Let $G = (V, E)$ be a connected graph. The *diameter* of G is

$$\text{diam}(G) = \max_{u, v \in V} d(u, v),$$

the greatest distance between any two vertices of G.

The trees on $n \geq 3$ vertices range in diameter from $2 = \text{diam}(S_n)$ to $n - 1 = \text{diam}(P_n)$. In Fig. 2.5, the trees on 6 vertices have been arranged in order of nonincreasing diameter.

Recall that the distance between two vertices in a connected graph is the length of a shortest path between them. If the graph is a tree, there is no need to search among various paths for the shortest one.

2.27 Theorem. *Suppose u and v are two (different) vertices of some tree T. Then there is a unique path in T from u to v.*

Proof. That there must be some path follows from the fact that T is connected. If P and Q are two paths from u to v then either they are identical or, as in the proof of Theorem 2.23, T would contain a cycle. ∎

Another obvious fact about the trees in Fig. 2.5 is that each of them has (at least) two vertices of degree 1.

2.28 Definition. Let G be a graph. A vertex of G of degree 1 is called a *pendant* vertex.

2.29 Lemma. *Let T be a tree on $n > 1$ vertices. Then T has two pendant vertices.*

Proof. Let u and v be vertices of T such that $d(u, v) = \text{diam}(T)$. If $d_T(v) > 1$, then v has a neighbor, w, that is not on the unique path from u to v. Because it passes through v, the path from u to w is longer than $\text{diam}(T)$. A similar contradiction is produced by assuming $d_T(u) > 1$. ∎

Among the applications of this lemma is the following.

2.30 Theorem. *If T is a tree on n vertices, then its chromatic polynomial is $p(T, x) = x(x - 1)^{n-1}$.*

Proof. The proof is by induction on n. If $n = 1$, then $p(T, x) = x$ and the proof is complete. If $n > 1$, let u be a pendant vertex of T and let $e = uw$ be the unique edge of T incident with u. Then $T - e$ is a disconnected graph having two components, one equal to the isolated vertex u and the other isomorphic to the tree T/e. By chromatic reduction and Equation (9),

$$p(T, x) = p(T - e, x) - p(T/e, x)$$
$$= xp(T/e, x) - p(T/e, x)$$
$$= (x - 1)p(T/e, x).$$

Because T/e is a tree on $n - 1$ vertices, the induction hypothesis gives $p(T/e, x) = x(x - 1)^{n-2}$. ∎

As we already know, the chromatic polynomial, by itself, is not a complete list of invariants. In fact, by Theorem 2.30, any two trees on n vertices are chromatically equivalent. What other shared properties require graphs to be chromatically equivalent? What does $p(G, x)$ tell us about the structure of G? Complete answers to these questions have not yet been obtained.

2.31 Corollary. *If T is a tree on $n > 1$ vertices, then $\chi(T) = 2$. Thus, every tree is a bipartite graph.*

The result is an immediate consequence of either Theorem 2.23 or Theorem 2.30.

Let G be the overlap of K_3 and K_2 in K_1. A single step of chromatic reduction leads to

$$\begin{array}{c}\circ\!\!-\!\!\circ\\ |\diagdown|\\ \circ\quad\circ\end{array} = \begin{array}{c}\circ\quad\circ\\ |\diagdown|\\ \circ\quad\circ\end{array} - \begin{array}{c}\circ\\ |\diagdown\\ \circ\quad\circ\end{array}.$$

From this expression and Theorem 2.30, it follows that

$$p(G, x) = x(x-1)^3 - x(x-1)^2$$
$$= x(x-1)^2(x-2).$$

The word "paraffin" may conjure up images of homemade jam topped off by a layer of wax. In chemistry, however, the word refers to a family of hydrocarbons with chemical formula C_nH_{2n+2}. Also known as *alkanes*, the paraffins are obtained from distilled petroleum. (Paraffin wax is a mixture of these compounds.) The *octanes*, C_8H_{18}, are a paraffin subfamily consisting of 18 *isomers*, not because there are 18 hydrogen atoms but because the eight-atom carbon skeleton can take any one of 18 different arrangements.[5] Five of the 18 octane carbon skeletons[6] are illustrated in Fig. 2.6.

Modern organic chemists have synthesized and/or isolated several million different molecules. Even more remarkable is their ability to predict chemical properties in advance of synthesis. Among the tools used in these predictions are various *topological indices*.[7] The first topological index was introduced by Harry Wiener of Brooklyn College, who established a remarkable correlation between it and boiling point (hence viscosity and surface tension) of the paraffins.[8]

---

[5] Because the valence of carbon is four (and because paraffins have no double carbon bonds), the arrangement of the hydrogen atoms is uniquely determined by the molecule's carbon skeleton.

[6] There are, in fact, 23 nonisomorphic trees on 8 vertices, but only 18 of them satisfy $\Delta(T) \leq 4$, a requirement imposed by the valence of carbon.

[7] A term coined by Haruo Hosoya in 1971.

[8] H. Wiener, *J. Amer. Chem. Soc.* **69** (1947), 17–20. Also see D. H. Rouvray, *Scientific American*, Sept. 1986, pp 40–47. Extensions of the Wiener index have been used in a variety of ways from predicting antibacterial activity in drugs [M. Medić-Sarić et al., *Acta Pharm. Jugosl.* **33** (1983), 199–208] to modeling crystalline phenomena [D. Bonchev et al., *Physica Status Solidi* **55** (1979), 181–187].

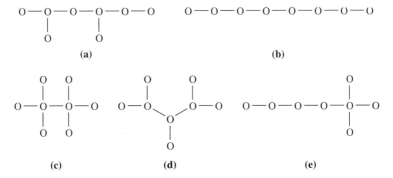

Figure 2.6. Five of the 18 isomers of octane.

2.32 Definition. Let $T = (V, E)$ be a tree with vertex set $V = \{v_1, v_2, \ldots, v_n\}$. The *Wiener index* of T is

$$W(T) = \sum_{i<j} d(v_i, v_j), \qquad (16)$$

the sum of the distances between all $C(n, 2)$ pairs of vertices of T.

2.33 Example. The trees in Fig. 2.6 have Wiener indices 71, 84, 58, 65, and 71, respectively. We shall have more to say about the Wiener index in Chapter 9. □

Various paraffins are used as fuel for internal combustion engines. In general, the more "branched" a paraffin molecule is, the less likely it will be to self-ignite or "knock." The more branched the molecule, the higher its "octane rating." In 1979, A. T. Balaban introduced a quantitative measure of branching for paraffin molecules.[9]

2.34 Balaban's Tree Pruning Algorithm. Suppose T is a tree on n vertices.

1. Let $G = T$ and $B = 0$.
2. If $n \leq 2$, then step 8.
3. $p =$ the number of pendant vertices of G.
4. $B = B + p^2$.
5. Replace G with the tree obtained by deleting its pendant vertices and the edges incident with them.
6. $n = n - p$.
7. Go to step 2.

---

[9] A.T. Balaban, Five new topological indices for the branching of tree-like graphs, *Theor. Chim. Acta* **53** (1979), 355–375.

Chap. 2 Chromatic Number

8. $B = B + n^2$.
9. $B(T) = B$ is the *Balaban centric index* of T.

In more conventional language, let $T_1 = T$ be a tree with $n \geq 3$ vertices, p_1 of which are pendants. Let T_2 be the tree obtained from T_1 by deleting all of its pendant vertices and the edges incident with them. (Then T_2 is the subtree of T_1 induced on its nonpendant vertices.) Suppose T_2 has p_2 pendant vertices. Let T_3 be the tree obtained from T_2 by deleting all p_2 of its pendants. Continuing this process, one eventually arrives at a tree T_r isomorphic either to K_2 or to K_1, in which case p_r is defined to be 2 or 1, respectively. The Balaban centric index is

$$B(T) = \sum_{i=1}^{r} p_i^2.$$

2.35 Example. Figure 2.7 illustrates the computation of a Balaban centric index. Tree $T = T_1$ has $p_1 = 5$ pendant vertices. Their removal produces the tree $T_2 \cong S_4$, having $p_2 = 3$ pendants. Deleting the three pendant vertices from T_2 yields $T_3 \cong K_1$, so $r = 3$, $p_3 = 1$, and $B(T) = 5^2 + 3^2 + 1^2 = 35$. □

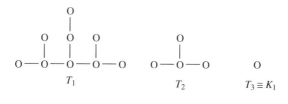

Figure 2.7

Suppose T is a tree of diameter d. At each iteration of Balaban's Tree Pruning Algorithm the diameter is reduced by 2. Therefore, $T_r \cong K_1$ when d is even, and $T_r \cong K_2$ when d is odd.

EXERCISES

1. The Hawaiian alphabet consists of the vowels a, e, i, o, u and the consonants h, k, l, m, n, p, and w.

    (a) Show that 20,736 different four-letter "words" (most of them nonsense words) can be constructed from the 12-letter Hawaiian alphabet.

    (b) Show that 456,976 different four-letter words can be constructed from the 26-letter English alphabet.[10]

---

[10] No more than 350,000 words are defined in a typical "unabridged" dictionary of the English language.

(c) How many four-letter words can be constructed from the Hawaiian alphabet if the first and third letters are consonants and the second and last letters are vowels?

2 Illustrate the following:
   (a) 6 (different) red–white–blue proper colorings of K_3.
   (b) 12 red–white–blue proper colorings of $P_3 = K_3 - e$.
   (c) 6 red–white–blue proper colorings of $K_2 = K_3/e$.

3 Let $G = K_2 \vee K_3^c$.
   (a) Illustrate G. (b) Compute $p(G, x)$.

4 Justify Equation
   (a) (5). (b) (6a). (c) (6b). (d) (7a). (e) (7b). (f) (8a). (g) (8b).

5 Prove that ω and α are (graph) invariants.

6 Let H be the overlap of K_3 and K_2 in K_1. Let G be an overlap of H and K_2 in K_1.
   (a) Show that there are exactly three nonisomorphic possibilities for G.
   (b) Show that the three nonisomorphic graphs from part (a) share the chromatic polynomial $x(x-1)^3(x-2)$.
   (c) Find a family of three nonisomorphic, chromatically equivalent graphs on 5 vertices.

7 Prove that the chromatic polynomial is an invariant.

8 Let $G = K_4 - e$.
   (a) Show that the characteristic polynomial of $A(G)$ is $x(x+1)(x^2 - x - 4)$.
   (b) Show that $\chi(G) \leq 1 + \rho(G)$, where $\rho(G)$ is the largest eigenvalue of $A(G)$.[11]

9 Compute the following:
   (a) $p(P_3, x)$. (b) $p(C_4, x)$. (c) $p(C_5, x)$.

10 Let G and H be graphs on disjoint sets of vertices. Suppose $d(G) = (d_1, d_2, \ldots, d_s)$ and $d(H) = (\delta_1, \delta_2, \ldots, \delta_t)$. Show that $d(G \vee H)$ is a rearrangement of the sequence
$$(d_1 + t, d_2 + t, \ldots, d_s + t, \delta_1 + s, \delta_2 + s, \ldots, \delta_t + s).$$

---

[11] It was proved by H. S. Wilf, [The eigenvalues of a graph and its chromatic number, *J. London Math. Soc.* **42** (1967), 330–332] that $\chi(G) \leq 1 + \rho(G)$ for all G.

Chap. 2 Chromatic Number

11 It follows from Theorem 2.30 that S_4 and P_4 are nonisomorphic, chromatically equivalent graphs. Find another pair of nonisomorphic, chromatically equivalent graphs on four vertices.

12 Let G be a graph with n vertices and k connected components. Prove the following:

(a) $p(G, x) = x^n - c_1 x^{n-1} + c_2 x^{n-2} - \cdots + (-1)^{n-k} c_{n-k} x^k$. (In other words, prove that $c_t = 0$ for $t > n - k$.)

(b) The coefficients of $p(G, x)$ alternate in sign, in other words, that $c_i > 0$, in part (a), $1 \le i \le n - k$.

(c) $p(G, t) \ne 0$ for every (real number) $t < 0$.

13 Suppose G^c is a disconnected graph with r components, $C_1, C_2, \ldots, C_r$.

(a) Prove that $\chi(G) = \sum \chi(C_i^c)$.

(b) Prove that $\omega(G) = \sum \omega(C_i^c)$.

14 Let T be a tree on n vertices, p of which are pendants. Suppose $\text{diam}(T) = d$. Prove the following:

(a) $n + 1 \ge p + d$. (b) $pd \ge 2(n - 1)$.

15 Compute the chromatic polynomial of the "broken wheel":

$$G = \text{[graph illustration]}$$

and express your answer in a form consistent with Equation (14).

16 Illustrate a graph G such that the following hold:

(a) $\chi(G) = \omega(G)$.

(b) $\chi(G) = \omega(G) + 1$.

(c) $\chi(G) = \omega(G) + 2$.

17 Compute the chromatic polynomial of the *wheel* $W_5 = $ [graph illustration].

18 In modern telecasts of National Football League games, one is frequently invited to examine important plays from the "reverse angle." Let's look at chromatic reduction from the reverse angle.

(a) Show that Equation (10) can be written in the form $p(G, x) = p(G + e, x) + p(G/e, x)$, where $G + e$ is obtained from G by adding a new edge $e = uv$ that was not there before, and G/e is obtained from G by identifying vertices u and v (and eliminating any multiple edges that may arise in the process).

(b) Show that the following picturesque example of this reverse angle approach is consistent with Equation (11):

$$\begin{matrix} o\!-\!o \\ |\diagup| \\ o\!-\!o \end{matrix} \;=\; \begin{matrix} o\!-\!o \\ |\!\times\!| \\ o\!-\!o \end{matrix} \;+\; \begin{matrix} o\!-\!o \\ |\diagup \\ o \end{matrix}$$

(c) If G is a graph on n vertices, prove that $p(G, x)$ is a nonnegative integer linear combination of the falling factorial functions $x^{(k)}, k \leq n$.

19 Use the reverse angle technique of Exercise 18 to compute the chromatic polynomial of the following:

(a) $G = K_5 - e$. (b) $G = (K_2 \oplus K_2 \oplus K_2)^c$.

20 Let H_1 and H_2 be graphs on disjoint sets of n_1 and n_2 vertices, respectively. A *coalescence* of H_1 and H_2 is a graph on $n_1 + n_2 - 1$ vertices that can be obtained from $H_1 \oplus H_2$ by identifying some vertex of H_1 with any vertex of H_2. Denote by $H_1 * H_2$ any one of the $n_1 n_2$ different coalescences of H_1 and H_2.

(a) Show that $p(H_1 * H_2, x) = p(H_1, x) p(H_2, x)/x$.

(b) Without actually computing it, prove that the graphs in Fig. 2.8 all have the same chromatic polynomial.

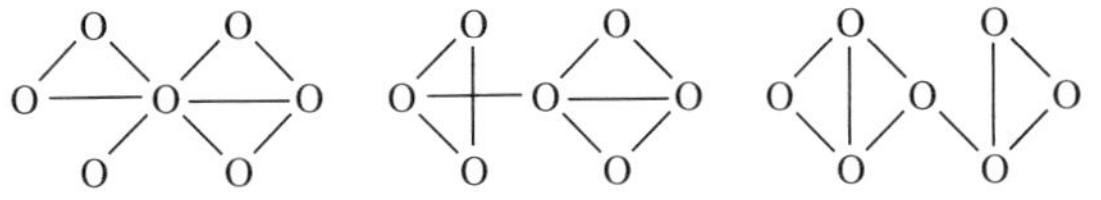

Figure 2.8

21 Without actually computing it, prove that the graphs in Fig. 2.9 all have the same chromatic polynomial.

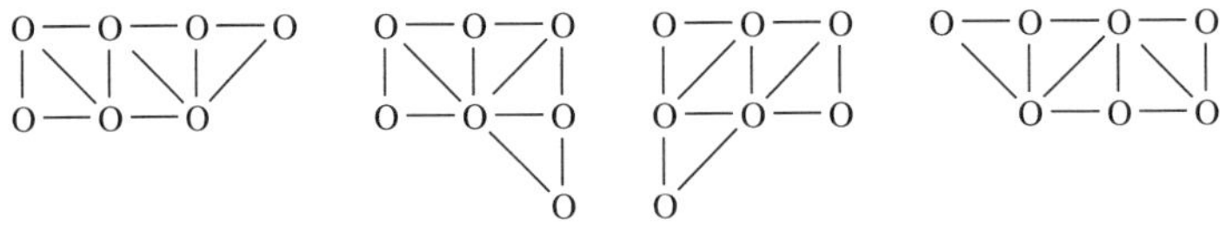

Figure 2.9

22 Suppose that $f(x)$ and $g(x)$ are defined in terms of falling factorial functions by

$$f(x) = \sum_{i=0}^{r} a_i x^{(i)} \quad \text{and} \quad g(x) = \sum_{j=0}^{s} b_j x^{(j)}.$$

Then the *join-product* of f and g is defined by

$$f(x) \vee g(x) = \sum_{k=0}^{r+s} \left(\sum_{t=0}^{k} a_t b_{k-t} \right) x^{(k)}.$$

For example, $(x^{(3)} + x^{(2)}) \vee (x^{(3)} + 2x^{(2)} + 5x^{(1)}) = x^{(6)} + 3x^{(5)} + 7x^{(4)} + 5x^{(3)}$. It turns out that the chromatic polynomial of a join of graphs is the join product of their chromatic polynomials. Thus, to compute $p(H_1 \vee H_2, x) = p(H_1, x) \vee p(H_2, x)$, all we need is to express $p(H_1, x)$ and $p(H_2, x)$ in terms of falling factorial functions. (See Exercise 18(c).) Use this approach to show that $p(P_3 \vee C_4, x) = x(x-1)(x-2)(x-3)(x^3 - 12x^2 + 50x - 71)$.

23 Use the join-product formula from Exercise 22 to compute the following:

(a) $p(K_1 \vee K_3^c, x)$. (b) $p(K_2^c \vee K_3^c, x)$. (c) $p(K_3^c \vee K_3^c, x)$.

24 Let G be a graph with n vertices, m edges, and chromatic polynomial $p(G, x) = x^n - c_1 x^{n-1} + \cdots$. Prove that $c_1 = m$.

25 Note that $G =$ is an "overlap" of two copies of C_4 in P_3. Without computing $p(G, x)$, show that it could not possibly be equal to $f(x) = p(C_4, x)^2 / p(P_3, x)$.

26 Prove that $p(C_n, x) = (x-1)^n + (-1)^n (x-1)$.

27 The *wheel* on $n+1$ vertices is $W_n = C_n \vee K_1$. (Exercise 17 involves W_5.) Compute $p(W_n, x)$.

28 Suppose G is a fixed but arbitrary graph. Prove or disprove that the roots of $p(G, x)$ are all real.

29 Prove that $f(x) = x^6 - 12x^5 + 54x^4 - 112x^3 + 105x^2 - 36x$ is not the chromatic polynomial of any graph.

30 Let $G = (V, E)$ be a graph with n vertices and m edges. Suppose $e = uv \in E$. To *subdivide* e means, informally, to put a new vertex in the middle of e. Of course, adding a vertex changes the graph. Let $H = (W, F)$ be the new graph. Then $W = V \cup \{w\}$, where $w \notin V$, and $F = (E \setminus \{uv\}) \cup \{uw, wv\}$ is the set obtained from E by replacing $e = uv$ with two new edges, uw and wv. In particular, $d_H(w) = 2$. If every edge of G is subdivided, the resulting graph, $S(G)$, has $n + m$ vertices and $2m$ edges. Prove that $S(G)$ is bipartite.

31 Prove that every tree on n vertices has $n-1$ edges.

32 If $p(G, x) = x^n - c_1 x^{n-1} + c_2 x^{n-2} - \cdots + (-)^{n-1} c_{n-1} x$, then G is both connected and bipartite if and only if c_{n-1} is odd. Use this criterion to prove that

(a) every tree is bipartite.

(b) C_n is bipartite if and only if n is even. (*Hint*: Exercise 26.)

33 Denote by t_n the number of nonisomorphic trees on n vertices.

(a) Prove that $t_4 = 2$.

(b) Illustrate the $t_5 = 3$ nonisomorphic trees on 5 vertices, explaining how you can be sure that they are nonisomorphic.

(c) Illustrate the $t_7 = 11$ nonisomorphic trees on 7 vertices.

(d) Illustrate the 11 nonisomorphic trees with 9 vertices and diameter 6.

34 A *cotree* is a graph whose complement is a tree.

(a) Illustrate the nonisomorphic cotrees on $n = 4$ vertices.

(b) Illustrate the nonisomorphic cotrees on $n = 5$ vertices.

(c) Let G be a disconnected cotree on $n \geq 4$ vertices. Prove that G has two components, one of which is an isolated vertex.

35 Let G be a connected graph. Prove or disprove that the diameter of G is equal to the length of a longest path in G.

36 Let G be a connected, antiregular graph on $n \geq 2$ vertices. (See Chapter 1, Exercise 20.) Prove the following:

(a) $G^c = H \oplus K_1$, where H is a connected, antiregular graph on $n - 1$ vertices.

(b) Up to isomorphism, there is exactly one connected, antiregular graph on n vertices.

37 Let A_n be the unique, connected antiregular graph on n vertices. (See Exercise 36(b).) Then $A_1 = K_1$, $A_2 = K_2 = P_2$, $A_3 = P_3 = S_3$, and so on. Prove the following:

(a) Up to isomorphism, $A_n \oplus K_1$ is the unique disconnected antiregular graph on $n + 1$ vertices.

(b) $A_{n+1} \cong A_n^c \vee K_1$, the join of A_n^c and K_1.

(c) $A_{n+1} \cong (A_{n-1} \oplus K_1) \vee K_1$, $n \geq 3$.

(d) $\omega(A_n) = \lfloor n/2 \rfloor + 1$.

(e) $\alpha(A_n) = \lfloor (n+1)/2 \rfloor$.

38 Let A_n be the unique, connected antiregular graph on $n \geq 2$ vertices. (See Exercise 36(b).) Prove the following:

(a) $p(A_n, x) = \begin{cases} x[(x-1)(x-2)\cdots(x-s)]^2, & n \text{ odd,} \\ x[(x-1)(x-2)\cdots(x-s+1)]^2(x-s), & n \text{ even,} \end{cases}$

where $s = \lfloor n/2 \rfloor$, the integer part of $n/2$.

(b) $\chi(A_n) = \lfloor n/2 \rfloor + 1$.

(c) $\chi(A_n)\chi(A_n^c) = \lfloor (n+1)^2/4 \rfloor$.

39 Let G be a connected graph having n vertices and $n-1$ edges. Prove that G is a tree.

40 Let G be a connected graph on $n \geq 2$ vertices. Prove that $\alpha(G) \leq \mathbf{i}(G)$, where $\mathbf{i}(G)$ is the intersection number introduced in Chapter 1, Exercise 31.

41 Let G be a connected graph of diameter D. Prove that $A(G)$ has at least $D+1$ distinct eigenvalues.

42 Show that the Wiener index of the path is $W(P_n) = C(n+1, 3)$.

43 Let T be a tree on k vertices. Let G be a graph with minimum vertex degree $\delta(G) \geq k-1$. Prove that T is isomorphic to a subgraph of G.

44 Show that the trees in Fig. 2.10 have the same Wiener index.

Figure 2.10

45 Let G be a connected graph. The *eccentricity* of $u \in V(G)$ is $\mathrm{ec}(u) = \max_{v \in V(G)} d(u, v)$, the maximum of the distances from u to the other vertices of G. The *radius* of G is $\mathrm{rad}(G) = \min_{u \in V(G)} \mathrm{ec}(u)$. The *center* of G is $\{w \in V(G): \mathrm{ec}(w) = \mathrm{rad}(G)\}$.

(a) Prove or disprove that the diameter of a connected graph G is equal to $2\,\mathrm{rad}(G)$.

(b) Prove that the center of a tree consists either of a single vertex or of two adjacent vertices.

46 Compute the Balaban centric index for each tree in Fig. 2.10.

47 Let $\gamma_1(G) \geq \gamma_2(G) \geq \cdots \geq \gamma_n(G)$ be the eigenvalues of $A(G)$. (Because $A(G)$ is a symmetric matrix, its characteristic roots are all real.)

(a) If G is bipartite, prove that $\gamma_i(G) + \gamma_{n-i+1}(G) = 0$, $1 \le i \le n$.[12]

(b) If G is bipartite and n is odd, prove that $A(G)$ is singular.

48 Let G be a graph on n vertices. Prove that $\chi(G)\chi(G^c) \ge n$.

49 Prove that the cube graph, Q_k, of Chapter 1 Exercise 33, is bipartite.

50 (Tutte) Let G be a fixed but arbitrary graph. Prove that $p(G, t) \ne 0$ for all $t \in (0, 1)$.

51 Let $u \in V(G)$ be a so-called *simplicial vertex* of G — that is a vertex with the property that $N_G(u)$ is a clique. Prove that $p(G, x) = (x - d)p(H, x)$, where $H = G - u$ is the graph obtained from G by deleting vertex u and all the edges incident with it, and $d = d_G(u)$.

52 If $k = \chi(G) > \chi(H)$ for every proper subgraph H of G, then G is *critically k-chromatic*.

(a) Show that C_n, $n = 2t + 1$, is critically 3-chromatic.

(b) Show that K_n, $n \ge 2$, is critically n-chromatic.

(c) If $\chi(G) = k$, show that G has a critically k-chromatic subgraph.

(d) If G is critically k-chromatic, prove that $\delta(G) \ge k - 1$.

(e) If G is a critically k-chromatic graph with n vertices and m edges, prove that $k \le 1 + 2m/n$.

(f) Prove that $\chi(G) \le 1 + \ell$ where ℓ is the length of a longest path in G. (*Hint*: Parts (c) and (d) and Exercise 43.)

53 (Mycielski) Let $G_1 = K_1$ and $G_2 = K_2$. If $G_k = (V_k, E_k)$, suppose $V_k = \{v_1, v_2, \ldots, v_n\}$. Let $V_{k+1} = V_k \cup \{u_1, u_2, \ldots, u_n, w\}$ and define $E_{k+1} = E_k \cup A_w \cup B_u$, where $A_w = \{wu_i : 1 \le i \le n\}$ and $B_u = \{xu_i : x \in V_k$ and $xv_i \in E_k, 1 \le i \le n\}$.

(a) Show that $G_3 \cong C_5$.

(b) Illustrate G_4.

(c) If G_k has n vertices and m edges, show that G_{k+1} has $2n + 1$ vertices and $3m + n$ edges, $k \ge 2$.

(d) Show that $\omega(G_k) = 2$, $k \ge 2$.

---

[12] It was shown by C.A. Coulson and G.S. Rushbrooke [Note on the method of molecular orbitals, *Proc. Cambridge Philose Soc.* **36** (1940), 193–200] that G is bipartite if and only if $\gamma_i(G) + \gamma_{n-i+1}(G) = 0$, $1 \le i \le n$, a result known to chemists as the "pairing theorem." D.M. Cvetković [Bihromaticnost i spektar grafa, Matematička Biblioteka (Beograd) **41** (1969), 193–194] later showed that a *connected* graph is bipartite if and only if $\gamma_1(G) + \gamma_n(G) = 0$.

(e) Show that $\chi(G_k) = k, k \geq 1$.

(f) Show that the difference, $\chi(G) - \omega(G)$, can be arbitrarily large.[13]

54 Denote by $t_n(D)$ the number of nonisomorphic trees on n vertices that have diameter D.

   (a) Compute $t_7(D)$, $2 \leq D \leq 6$. (*Hint*: Exercise 33.)

   (b) Prove that $t_n(n-2) = t_n(3)$, $n \geq 4$.

   (c) Prove or disprove that $t_n(n-r) = t_n(r+1)$, $1 \leq r \leq n-3$.

---

[13] It was shown by H.A. Kierstead and J.H. Schmerl [Some applications of Vizing's theorem to vertex colorings of graphs, *Discrete Math.* **45** (1983), 277–285] that if G does not contain $K_5 - e$ or $K_{1,3}$ as an induced subgraph, then $\omega(G) \leq \chi(G) \leq \omega(G) + 1$.

3

Connectivity

Recall that $G \neq K_1$ is connected if every pair of its vertices is joined by a path. Both P_6 and C_6 satisfy this criterion; yet, while two vertices of P_6 are joined by a unique path, any two vertices of C_6 are joined by two different paths. It seems that C_6 is *more* connected than P_6. The main thrust of the chapter is to explore and quantify this notion of relative connectedness. It is not surprising that cycles should have a central place in the discussion. The surprise is that the investigation should be so demanding, leading as it does to some of the deepest theorems of the book. The payoff for this hard work comes in later chapters where applications can be as delightful as they are unexpected.

Suppose our interest in graph $G = (V, E)$ extends no further than some subset $W \subset V$. In that case, it may be useful to focus on a subgraph of the form $H = (W, F)$. This restriction naturally precludes certain edges of G from belonging to $F = E(H)$. In fact, the unique maximal subgraph of G that satisfies $V(H) = W$ is the induced subgraph $G[W]$ corresponding to $F = E \cap W^{(2)}$.

Alternatively, it may happen that our interest is drawn to some subset $F \subset E$. In this case, focusing on a subgraph $H = (W, F)$ does not preclude any vertices of G from belonging to $W = V(H)$. The unique maximal subgraph of G that satisfies $E(H) = F$ is $H = (V, F)$. Such a subgraph is said to *span* G. Thus, $H = (W, F)$ is a *spanning* subgraph of G if (and only if) $W = V(G)$. In some cases (e.g., chromatic reduction), attention is directed to F by default, the edges in $D = E \backslash F$ having been removed. From this perspective, H is a spanning subgraph of G if and only if there is a subset $D = \{e_1, e_2, \ldots, e_t\} \subset E$ such that

$$H = (V, E \backslash D)$$
$$= G - D$$
$$= G - e_1 - e_2 - \cdots - e_t. \tag{17}$$

3.1 Definition. Let G be a graph on $n \geq 2$ vertices. A *disconnecting set* of edges is a subset $D \subset E(G)$ such that $G - D$ is disconnected. The *edge connectivity*, $\varepsilon(G)$, is the smallest number of edges in any disconnecting set.

We adopt the convention that $\varepsilon(K_1) = 0$. Thus, $\varepsilon(G) = 0$ if and only if $G = K_1$ or G is disconnected. If $\varepsilon(G) = 1$, then G is a connected graph having an edge

e such that $G - e$ is disconnected. An edge whose removal increases the number of components is called a *cut-edge* of G.

3.2 Lemma. *If G is a graph, then $\varepsilon(G) \leq \delta(G)$. That is, the edge connectivity of G can be no larger than the smallest of its vertex degrees.*

Proof. If $\varepsilon(G) = 0$ there is nothing to prove. So, let G be a connected graph on $n \geq 2$ vertices and suppose $u \in V(G)$ is a vertex of degree $d(u) = \delta(G) > 0$. Since $D = \{uv : v \in N_G(u)\}$ is a disconnecting set of edges, $\delta(G) = o(D) \geq \varepsilon(G)$. ∎

3.3 Theorem. *Suppose G is a connected graph. Let $e = uv \in E(G)$. Then e is a cut-edge of G if and only if $P = [u, v]$ is the only path in G from u to v.*

Theorem 3.3 may seem obvious because it is consistent with intuitive notions about connectivity. While intuition can be a source of inspiration, additional tools are generally needed to forge rigorous proofs. Among the most powerful of these tools are the definitions. Indeed, mathematical definitions *exist* to be used as tools in building proofs.[1]

Proof. If $[u, v]$ is the only path in G from u to v, then there is no path in $G - e$ from u to v and, by definition, the spanning subgraph $G - e$ is disconnected.

Conversely, let $e = uv$ be a cut-edge of G. Suppose H_1 and H_2 are two components of $G - e$. Let $w_i \in H_i, i = 1, 2$. Without loss of generality, we may assume that $u \notin H_2$. Let $Q = [u = x_0, x_1, \ldots, x_r = w_2]$ be a path in G from u to w_2. Because Q is not a path in $G - e$, edge e must lie along Q. Since the x's are distinct and $u = x_0$, it can only be that $e = \{x_0, x_1\}$. Therefore, $x_1 = v$ and $Q' = [v = x_1, x_2, \ldots, x_r = w_2]$ is a path in $G - e$ from v to w_2. Hence, $v \in H_2$. A similar argument now shows that $u \in H_1$. Because u and v are in different components of $G - e$, edge e must lie along every path in G from u to v. If $P = [u = y_0, y_1, \ldots, y_k = v]$ is such a path, then, since the y's are distinct and $e = uv$ is an edge of P, we have $v = y_1$; that is, $k = 1$ and $P = [u, v]$. ∎

It is because of Theorem 3.3 that a cut-edge is sometimes called a *bridge*. (See Fig. 3.1.) If $\varepsilon(G) = 1$, then G is barely connected in the sense that it can be disconnected by the removal of a single, well-chosen edge. If T is a tree, then any edge will do. *Every* edge of T is a bridge.

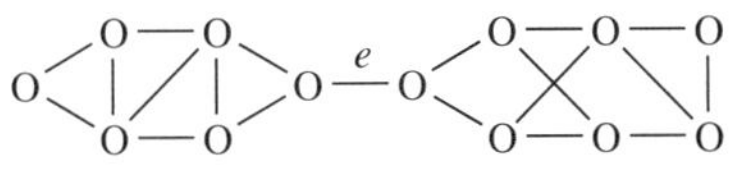

Figure 3.1

---

[1] Perhaps because he did not understand this point, the Roman writer Favorinus commented in his *Miscellaneous History* that Pythagoras "used definitions throughout the subject matter of mathematics."

3.4 Corollary. *Suppose G is a connected graph. Let $e = uv \in E(G)$. Then e is a bridge if and only if no cycle of G contains both u and v.*

The proof is left to the exercises.

If G is a connected graph without bridges, Corollary 3.4 implies that every edge of G must lie on a cycle.

3.5 Definition. Let $G = (V, E)$ be a graph and suppose $u \in V$. The *vertex deleted* graph $G - u = G[V \setminus \{u\}]$.

If $D = \{uv \in E : v \in N_G(u)\}$, then $G - u = (W, F)$, where $W = V \setminus \{u\}$ and $F = E \setminus D$. Deleting vertices is a more invasive kind of surgery than deleting edges. When an edge is deleted, the vertices incident with it remain in place (which is why the result is a spanning subgraph). When a vertex is deleted, all the edges incident with it are removed as well (which is why $G - u$ is a graph). Denote by

$$G - S = G - v_1 - v_2 - \cdots - v_r \tag{18}$$

the graph obtained from G by deleting the vertices in $S = \{v_1, v_2, \ldots, v_r\} \subset V(G)$, that is, $G - S = G[V \setminus S]$.

3.6 Definition. Suppose $K_n \neq G$ is a graph. A *vertex cut* (or *separating set*) of G is a subset $S \subset V(G)$ such that $G - S$ is disconnected. The *connectivity*, $\kappa(G)$, is the smallest number of vertices in any vertex cut of G.

Complete graphs were exempted from Definition 3.6 because they have no vertex cuts. They are the only such graphs. If u and w are nonadjacent vertices of G, then $S = V \setminus \{u, w\}$ is a vertex cut. Definition 3.6 is extended to complete graphs by defining $\kappa(K_n) = n - 1$. Thus, $\kappa(G) = 0$ if and only if $G = K_1$ or G is disconnected; that is, $\kappa(G) = 0$ if and only if $\varepsilon(G) = 0$.

If $\kappa(G) = 1$, then either $G = K_2$ or G is a connected graph with a vertex v such that $G - v$ is disconnected. A vertex whose removal increases the number of components of G is called a *cut-vertex* (or *point of articulation*). The graph illustrated in Fig. 3.1 has two points of articulation, namely the vertices incident with edge e. The fact that $\kappa(G)$ is called the connectivity of G (rather than its "vertex connectivity") reflects the view that, as a quantitative measure of connectivity, $\kappa(G)$ is more important than $\varepsilon(G)$.

3.7 Theorem. *If G is a graph, then $\kappa(G) \leq \varepsilon(G)$.*

Proof. We have already seen that $\varepsilon(G) = 0$ if and only if $\kappa(G) = 0$. Assume that $\kappa(G) \leq \varepsilon(G)$ whenever $\varepsilon(G) = k \geq 0$, and suppose G is a graph with $\varepsilon(G) = k + 1$. Let $D \subset E(G)$ be a minimum disconnecting set. Then $o(D) = k + 1$ and $G - D$ is disconnected. Suppose $e \in D$ is fixed but arbitrary and let $G_e = G - e$. Then $\varepsilon(G_e) = k$. By the induction hypothesis, $\kappa(G_e) \leq k$. Let $S \subset V(G_e) = V(G)$ be a minimum vertex cut of G_e. Then $o(S) \leq k$ and $G_e - S$ is disconnected. If

$G_S = G - S$ is disconnected, the proof is finished. If G_S is connected, then e cannot be incident in G with any vertex of S. Hence, $G_S - e = G - S - e = G - e - S = G_e - S$ is disconnected; that is, e is a cut-edge of G_S. If either vertex of e has degree greater than one in G_S, it is a cut-vertex of G_S, proving that $\kappa(G) \leq k + 1 = \varepsilon(G)$. Otherwise, because it is connected, $G_S \cong K_2$, meaning that $n - 2 = o(S) \leq k$. Therefore, using Lemma 3.2, $n - 1 \leq k + 1 = \varepsilon(G) \leq \delta(G) \leq n - 1$. But, $\delta(G) = n - 1$ if and only if $G = K_n$, in which case $\kappa(G) = n - 1 = \varepsilon(G)$. ∎

3.8 Definition. A *block* is a connected graph with no cut-vertices.

If $\kappa(G) \geq 2$, then G is a block. The converse fails for two reasons: K_1 and K_2. If $n = o(V(G)) \geq 3$, then G is a block if and only if $\kappa(G) \geq 2$.

Every complete graph is a block as is $C_n, n \geq 3$. If T is a tree on $n \geq 3$ vertices, it must contain at least one vertex u of degree $d_T(u) \geq 2$. Because $T - u$ is disconnected, T is not a block. Indeed, by Theorem 3.7, $1 = \varepsilon(T) \geq \kappa(T) \geq 1$, for any tree T on $n \geq 2$ vertices.

3.9 Definition. A maximal connected subgraph of G that has no cut-vertices is a *block of G*.

Suppose B is a block of G and H is a subgraph of G. In the context of Definition 3.9, "maximal" means that if B is a proper subgraph[2] of H, then either H is disconnected or it has a cut-vertex.

The graph illustrated in Fig. 3.1 has the three blocks shown in Fig. 3.2. Note that a subgraph of G can be a block *without* being a block of G. Because neither has a cut-vertex, the complete graphs K_1 and K_2 are blocks. If B is a block of G, then $B \cong K_1$ if and only if B corresponds to an isolated vertex, and $B \cong K_2$ if and only if it corresponds to a bridge. If $\kappa(G) \geq 2$, then, while it may have many vertices and edges, the only block of G is G itself.

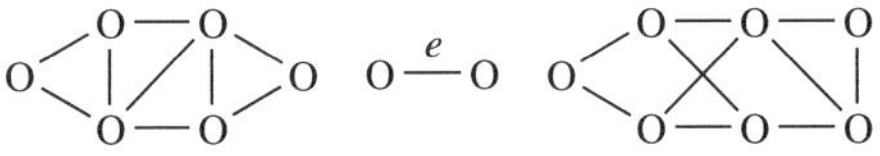

Figure 3.2

3.10 Theorem. *Suppose $H = (W, F)$ is a connected subgraph of $G = (V, E)$.*

(i) *If H has no cut-vertices, then $G[W]$ is connected and it has no cut-vertices.*
(ii) *The blocks of G are induced subgraphs.*
(iii) *If H has no cut-vertices, then H is a subgraph of some block of G.*

---
[2] A subgraph B of H is *proper* if $B \neq H$.

Proof. Because H is a subgraph of $G[W]$, every path in H is a path in $G[W]$. Moreover, for any $w \in W$, every path in $H - w$ is a path in $G[W] - w$. This proves part (i).

If H is a block of G, then, by maximality and part (i), $H = G[W]$, proving part (ii). If $G[W]$ is a block, then part (iii) follows from part (i). Otherwise, it must be that $G[W]$ fails to be maximal, so there is a connected subgraph $H_1 = (W_1, F_1)$ of G such that H_1 has no cut-vertices, $G[W]$ is a subgraph of H_1, and $o(W_1) > o(W)$. By part (i), we may assume $H_1 = G[W_1]$. If H_1 is not a block, then, by the same reasoning, there is a connected subgraph $H_2 = G[W_2]$, with no cut-vertices, such that W_1 is a proper subset of W_2. Because $o(W) < o(W_1) < o(W_2) < \cdots \leq o(V)$, this process ends at a block of G. ∎

While a block B of G has no cut-vertices of its own, B may contain cut-vertices of G. (See Figs. 3.1 and 3.2.)

3.11 Corollary. *Let $G[X]$ and $G[Y]$ be two (different) blocks of the connected graph G. Then either $X \cap Y = \phi$ or $X \cap Y = \{z\}$, in which case z is a cut-vertex of G.*

Proof. Suppose $X \cap Y \neq \phi$. By the maximality of blocks, neither X nor Y is a proper subset of the other. Therefore, $G[X]$ and $G[Y]$ are proper subgraphs of $H = G[X \cup Y]$. Suppose u and w are different vertices of H. If u and w are both in X, or both in Y, there is a path in $G[X]$ or $G[Y]$ from u to w. Otherwise, there is a path in H from u to w through $X \cap Y$. Thus, H is connected. If $o(X \cap Y) \geq 2$, a similar examination of cases shows that H has no cut-vertex. So, from Theorem 3.10, H is contained in a block of G, contradicting the maximality of blocks $G[X]$ and $G[Y]$. Hence, $o(X \cap Y) = 1$.

Suppose $X \cap Y = \{z\}$. Let $x \in X$ and $y \in Y$ be vertices that are adjacent in G to z. Suppose there is a path P, in $G - z$, from x to y. Then $C = \langle P, z \rangle$ is a cycle of G containing x, y, and z. Because the connected subgraph C of G has no cut-vertices, it follows from Theorem 3.10 that C is a subgraph of some block $B = G[W]$. Since $y \in W \setminus X$, $B \neq G[X]$. So, $G[X]$ and $G[W]$ are different blocks of G. Because $x, z \in X \cap W$, $o(X \cap W) > 1$, contradicting the first part of the proof. Hence, there can be no path in $G - z$ from x to y, meaning that x and y belong to different components of $G - z$, meaning that z is a cut-vertex of G. ∎

3.12 Example. Each graph in Fig. 3.3 has the same four blocks, namely, $B_1 \cong C_4$, $B_2 \cong C_3 \cong K_3$, $B_3 \cong K_2 \cong P_2$, and $B_4 \cong K_4 - e$. Note, however, that while G_1 and G_2 each have three cut-vertices, G_3 has only two. □

3.13 Theorem. *Let G be a connected graph with exactly t blocks, $B_1, B_2, \ldots, B_t$. Then the chromatic polynomial*

$$p(G, x) = \frac{1}{x^{t-1}} \prod_{i=1}^{t} p(B_i, x).$$

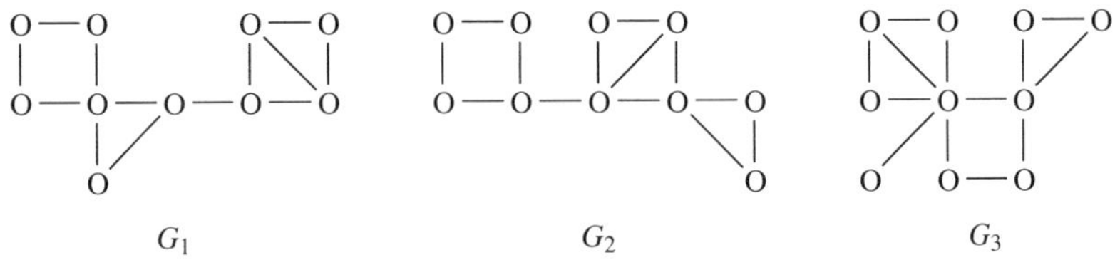

$$G_1 \qquad G_2 \qquad G_3$$

Figure 3.3

Proof. The result follows from Theorem 2.13 and the fact (see Exercise 13) that G can be "constructed" from its blocks in a sequence of steps, each of which is comprised of an overlap of two graphs in K_1. ■

3.14 Definition. Let G be a graph and let k be a positive integer. If $\kappa(G) \geq k$, then G is said to be k-connected.

If G is $(k+1)$-connected, then G is k-connected. A graph on $n \geq 2$ vertices is 1-connected if and only if it is connected, and a graph on $n \geq 3$ vertices is 2-connected if and only if it is a block.

3.15 Theorem. *Let G be a graph on $n \geq 3$ vertices. Then G is 2-connected if and only if, for all $u, w \in V(G)$, $u \neq w$, there exists a cycle of G containing both u and w.*

Proof. Suppose $\kappa(G) \geq 2$. Let $u \in V(G)$ be fixed but arbitrary and denote by $S \subset V(G)\setminus\{u\}$ the set of vertices that do not share a cycle with u. We will prove necessity by showing that $S = \phi$. So, suppose $w \in S$. Because $2 \leq \kappa(G) \leq \varepsilon(G)$, G has no cut-edges. Therefore (Corollary 3.4), every edge of G lies on a cycle, proving that $S \cap N_G(u) = \phi$. Let $P = [u = v_0, v_1, \ldots, v_{r+1} = w]$ be a path in G from u to w. If i is the smallest positive integer such that $v_i \in S$, then $i > 1$ because $v_1 \in N_G(u)$. Since the only restriction we have placed on w is that it belong to S, we may assume $w = v_i$, in other words, that $i = r+1$. Because $v_r \notin S$, there exists a cycle C containing both u and v_r.

Let $Q = [u = x_0, x_1, \ldots, x_t = w]$ be a path in $G - v_r$ from u to w. If the only vertex common to Q and C is u, then (see Fig. 3.4(a)) we can go along either subpath of C from u to v_r, cross over to Q by means of edge $v_r w$, and then complete a cycle by going backwards along Q to u. The existence of this new cycle contradicts $w \in S$. So, there must be a largest integer j such that $u \neq x_j \in C$. Because Q is a path in $G - v_r$, $x_j \neq v_r$. Thus (see Fig. 3.4(b)), we can go from v_r to x_j along the subpath of C that contains u, then along Q from x_j to w, and complete a cycle by crossing edge wv_r. Because both u and w lie on this new cycle, the assumption $w \in S$ is again contradicted. This final contradiction proves that $S = \phi$.

Conversely, suppose $v \in V(G)$ is fixed but arbitrary. Let u and w be two (different) vertices of $V(G)\setminus\{v\}$. Suppose $C = \langle u = x_1, \ldots, x_s = w, \ldots, x_t \rangle$ is

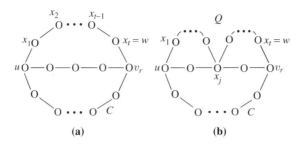

Figure 3.4

a cycle of G containing u and w. Then C contains two subpaths from u to w, namely, $P = [u = x_1, x_2, \ldots, x_s = w]$ and $Q = [u = x_1, x_t, \ldots, x_{s+1}, x_s = w]$. Because $n \geq 3$, and v was arbitrary, the existence of either path proves that G is connected. While P and Q overlap in u and w, having no other vertices in common, they are "internally disjoint." In particular, at least one of them is a path in $G - v$, proving that v is not a cut-vertex of G. ∎

Among the many applications of graphs is to the modeling of social interactions. In this context, descriptives like "isolated" and "neighbor" seem especially appropriate. Central to most sociological models is the network of linkages between nonadjacent vertices. Consider, for example, the graph G in Fig. 3.5. While the distances $d(x, y) = 2 = d(x, z)$, vertex x seems better connected to vertex z than to vertex y, at least in the sense of sociological networking. Let's try to quantify this notion of "networking connectivity."

Figure 3.5

If vertices u and v lie in different components of G, they are totally disconnected from each other, and there is little more to be said. The interesting case is when u and v lie in the same component. Thus, we may as well assume, right from the beginning, that G is connected.

3.16 Definition. Suppose u and v are nonadjacent vertices of a connected graph G. Then $S \subset V(G)$ *separates u and v* if they lie in different components of $G - S$. Let $\kappa_G(u, v)$ be the minimum number of vertices whose deletion separates u and v; that is, $\kappa_G(u, v) = \min o(S)$, where the minimum is over the subsets S of $V(G)$ that separate u and v.

The reader may confirm that $\kappa_G(x, y) = 2$ and $\kappa_G(x, z) = 3$ in Fig. 3.5. Because a set of vertices that separates u and v is a separating set of G, we obtain

$$\kappa_G(u, v) \geq \kappa(G). \tag{19}$$

Indeed, if $K_n \neq G$ is connected, then $\kappa(G) = \min \kappa_G(u, v)$, where the minimum is over all nonadjacent pairs $u, v \in V(G)$.

Another natural measure of networking connectivity is the following.

3.17 Definition. Suppose u and v are nonadjacent vertices of a connected graph G. Two paths from u to v are *internally disjoint* if, apart from u and v, no vertex of G lies on both paths. Let $\psi_G(u, v)$ be the maximum number of internally disjoint paths in G from u to v.

Suppose $P = [u, x_1, \ldots, x_r, v]$ and $Q = [u, y_1, \ldots, y_s, v]$ are internally disjoint paths in G from u to v. By definition of "path," the x's are all different from each other, and no two y's are equal. "Internally disjoint" means that

$$\{x_i : 1 \leq i \leq r\} \cap \{y_j : 1 \leq j \leq s\} = \phi.$$

In Fig. 3.5, $\psi_G(x, y) = 2$ and $\psi_G(x, z) = 3$.

It is worth emphasizing that the definitions of $\kappa_G(u, v)$ and $\psi_G(u, v)$ apply to *nonadjacent* vertices of a *connected* graph G. When G is understood, we may write $\kappa(u, v)$ in place of $\kappa_G(u, v)$ and set $\psi(u, v) = \psi_G(u, v)$. The obvious question, of course, is whether these two new measures of networking connectivity are related to each other. In Fig. 3.5, $\kappa(x, y) = \psi(x, y)$, and $\kappa(x, z) = \psi(x, z)$. Is that typical or just the result of a poorly chosen example?

3.18 Menger's Theorem. *If u and v are nonadjacent vertices of a connected graph G, then $\kappa(u, v) = \psi(u, v)$; that is, the minimum number of vertices whose deletion separates u and v is equal to the maximum number of internally disjoint paths in G from u to v.*

Proof. If $S \subset V(G)$ separates u and v, then every path from u to v must contain a vertex of S. Because no two internally disjoint uv-paths can contain the same vertex of S, $o(S) \geq \psi(u, v)$. Because S was arbitrary, $\kappa(u, v) \geq \psi(u, v)$.

The proof that $\psi(u, v) \geq \kappa(u, v)$ is more difficult. If it is false, then there is a counterexample, a connected graph G, and two vertices $u, v \in V(G)$ such that $uv \notin E(G)$ and $\kappa(u, v) > \psi(u, v)$. Let k be the smallest value of $\kappa(u, v)$ among all counterexamples. Because $\psi(u, v) \geq 1$, $k \geq 2$.

Among the counterexamples G, u, v with $\kappa(u, v) = k$, consider those with a minimum value of $n = o(V(G))$. Finally, from among these counterexamples, choose one with a minimum value of $m = o(E(G))$. (Note that m may not be the smallest number of edges among all counterexamples.) The remainder of the proof depends upon three observations about our minimum counterexample.

Chap. 3 Connectivity

Observation 1: The distance $d(u, v) > 2$.

Because $uv \notin E(G), d(u, v) \geq 2$. If $d(u, v) = 2$, there exists a vertex $w \in V(G)$ such that $[u, w, v]$ is a path in G. It follows that w is an element of any set that separates u and v. Let $H = G - w$. Because $k = \kappa_G(u, v) \geq 2, u$ and v belong to the same component, C, of H. Because of the way G was chosen, C is not a counterexample. Thus,

$$\kappa_G(u, v) - 1 = \kappa_C(u, v)$$
$$= \psi_C(u, v). \qquad (20)$$

Suppose $\{P_1, P_2, \ldots, P_\psi\}, \psi = \psi_G(u, v)$, is a family of internally disjoint paths in G from u to v. If w were on none of these paths, we could add $[u, w, v]$ and obtain a larger family, contradicting the definition of ψ. Because the P's are internally disjoint, w lies on exactly one of them. Hence, $\psi_C(u, v) = \psi_G(u, v) - 1$, and Equation (20) contradicts the fact that G is a counterexample.

Observation 2: If $S = \{w_1, w_2, \ldots, w_k\}$ separates u and v, then every vertex of S is adjacent to u or every vertex of S is adjacent to v. (Don't misunderstand. This observation must hold only for our hypothetical minimal counterexample.)

Because (Observation 1) $d(u, v) > 2$, no $w \in V(G)$ can be adjacent to both u and v. If all the w's are adjacent to u, we are finished. Otherwise, denote by C_u the component of $G - S$ containing u, and let $G[W]$ be the subgraph of G induced on the set of vertices $W = S \cup V(C_u)$. (See Fig. 3.6.) Define H to be the graph

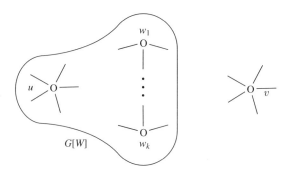

Figure 3.6

obtained from $G[W]$ by adding a new vertex v^* and new edges $\{v^*, w_i\}, 1 \leq i \leq k$. Observe that $\kappa_H(u, v^*) = k$. If H has $n = o(V(G))$ vertices, then $H \cong G$, and Observation 2 is proved. Otherwise, having fewer than n vertices, H is not a counterexample; that is, $\psi_H(u, v^*) = \kappa_H(u, v^*) = k$, so there are k internally disjoint paths in H from u to v^*. It follows that H (hence G) contains paths P_i from u to w_i, $1 \leq i \leq k$, that overlap only in u. Turning the argument around, there must also exist paths Q_i from v to w_i, $1 \leq i \leq k$, that overlap only in v. Denote by Q_i^{-1} the path in G from w_i to v that reverses Q_i. Overlapping P_i and Q_i^{-1} at w_i, $1 \leq i \leq k$, produces k internally disjoint paths in G from u to v. This contradicts the fact that G is a counterexample, and it finishes the proof of Observation 2.

Observation 3: Let $P = [u, x, y, \ldots, v]$ be a shortest path from u to v in our minimal counterexample G. (By Observation 1, $u, x, y,$ and v are all different vertices of G.) Then there exists a set $T \subset V(G)$ such that $\mathrm{o}(T) = k - 1$, and both $S_1 = T \cup \{x\}$ and $S_2 = T \cup \{y\}$ separate u and v.

Let $e = xy$. Because $\kappa_G(u, v) = k \geq 2$, u and v are in the same component of $G - x$. Hence, they are in the same component, H, of $G - e$. Let $\phi \neq T \subset V(H)$ be a minimum separator of u and v in H. Because H is not a counterexample, we have

$$\begin{aligned}
\mathrm{o}(T) &= \kappa_H(u, v) \\
&= \psi_H(u, v) \\
&\leq \psi_G(u, v) \\
&< \kappa_G(u, v) \\
&= k.
\end{aligned}$$

Therefore, T does not separate u and v in G. Evidently, $e = xy$ is a bridge in $G - T$. Thus, both $T \cup \{x\}$ and $T \cup \{y\}$ separate u and v in G. In particular, $\mathrm{o}(T) + 1 \geq k > \mathrm{o}(T)$, implying that $\mathrm{o}(T) = k - 1$. This completes the proof of Observation 3.

Because (Observation 3) $S_1 = T \cup \{x\}$ is a minimum separator of u and v in G, it follows from $ux \in E(G)$ and Observation 2 that $uw \in E(G)$, for all $w \in T$. Because $S_2 = T \cup \{y\}$ is a minimum separator of u and v, every vertex of S_2 is adjacent either to u or to v. Since we have already established that $uw \in E(G)$, for all $w \in T \neq \phi$, every vertex of S_2 must be adjacent to u. In particular, $uy \in E(G)$, contradicting the fact that $P = [u, x, y, \ldots, v]$ is a shortest path in G from u to v. This final contradiction finishes the proof of Menger's Theorem by showing that there are no counterexamples. ∎

3.19 Corollary. *Suppose G is k-connected. Let $w, u_1, u_2, \ldots, u_k$ be $k + 1$ (different) vertices of G. Then there exist paths in G from w to u_i, $1 \leq i \leq k$, no two of which overlap except in w.*

Proof. Let H be the graph obtained from G by adding a new vertex v^* and k new edges, $u_i v^*$, $1 \leq i \leq k$. From Exercise 8, H is k-connected. Therefore,

$$k \leq \kappa(H) \leq \kappa_H(w, v^*) \leq k. \tag{21}$$

By Menger's Theorem and (21), there are $\psi_H(w, v^*) = \kappa_H(w, v^*) = k$ internally disjoint paths in H from w to v^*. From the definition of H, these internally disjoint paths must be of the form $P_i = [w, \ldots, u_i, v^*]$, $1 \leq i \leq k$. Let $Q_i = [w, \ldots, u_i]$ be the path obtained from P_i by deleting vertex v^*, $1 \leq i \leq k$. Then $Q_1, Q_2, \ldots, Q_k$ are the paths we seek. ∎

3.20 Theorem. *Suppose $k \geq 2$ is an integer. Let G be a k-connected graph on n vertices. If U is a fixed but arbitrary k-vertex subset of $V(G)$, then there exists a cycle of G that contains every vertex of U.*

Proof. If $k = 2$, the result is half of Theorem 3.15. So, suppose $k \geq 3$ and let $U = \{u_1, u_2, \ldots, u_k\}$. Because G is 2-connected, there exists a cycle of G containing any given pair of u's. Suppose C is a cycle of G that contains a *largest* number, say r, of the vertices of U. If $r = k$, the proof is finished. Otherwise, $2 \leq r < k$, and we can renumber the elements of U, if necessary, so that it is the first r u's that belong to C and so that they occur consecutively around the cycle.

If every vertex of C belongs to U, then C is the cycle $\langle u_1, u_2, \ldots, u_r \rangle$. Let $w = u_{r+1}$. By Corollary 3.19, there exist paths P_i from w to u_i, $1 \leq i \leq r$, any two of which overlap only in w. In particular, P_i overlaps C only in u_i, $1 \leq i \leq r$. Let C' be the cycle obtained from C when edge $[u_1, u_2]$ is replaced by $P_1^{-1} P_2$. Then C' contains more than r elements of U, contradicting the maximality of C. Thus, C must contain a vertex $v \notin U$.

Because $k \geq r + 1$, Corollary 3.19 implies the existence of paths Q_i, $1 \leq i \leq r + 1$, such that Q_i is a path from w to u_i, $1 \leq i \leq r$, Q_{r+1} is a path from w to v, and any two (different) Q's overlap only in w. For each i, let v_i be the *first* vertex of Q_i that lies on C. (It may happen that $v_i = u_i$ for some $i \leq r$, and/or that $v_{r+1} = v$.) Denote by P_i the subpath of Q_i from w to v_i, $1 \leq i \leq r + 1$.

For each $i < r$, let S_i be the set of vertices in the subpath of C from vertex u_i up to, but not including, vertex u_{i+1}. Then S_i is analogous to a half-open interval $[u_i, u_{i+1})$. Let S_r be the corresponding interval along C from u_r to u_1. Then $u_i \in S_i$, $1 \leq i \leq r$, and the set of vertices comprising C is $S_1 \cup S_2 \cup \cdots \cup S_r$. Because, the distinct vertices $v_1, v_2, \ldots, v_{r+1}$ all belong to this disjoint union there exists a pair of them, say v_s and v_t, belonging to the same interval, say S_j. Without loss of generality, we may assume that $v_s = u_j$ or v_s lies on the subpath of C from u_j to v_t. (See Fig. 3.7.) Let C^* be the cycle obtained from C by replacing the path along C from v_s to v_t with $P_s^{-1} P_t$. Then C^* is a cycle of C containing $r + 1$ elements of U, contradicting the maximality of C and proving that r cannot be less than k. ∎

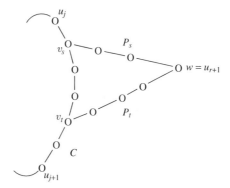

Figure 3.7

3.21 Example. Let G be the Petersen graph illustrated in Fig. 3.8. By Lemma 3.2 and Theorem 3.7, $3 = \delta(G) \geq \varepsilon(G) \geq \kappa(G)$. Because (Exercise 37) $\kappa(G) > 2$, it must be the case that $\varepsilon(G) = \kappa(G) = 3$. Since $\kappa(G) \leq \kappa(u, v) = \psi(u, v) \leq d_G(u)$, it follows that $\kappa(u, v) = 3 = \psi(u, v)$ for every pair of nonadjacent vertices u and v; for example, vertices 1 and 6 are joined by $P_1 = [1, 2, 6]$, $P_2 = [1, 10, 7, 6]$, and $P_3 = [1, 9, 5, 6]$. By Corollary 3.19, vertex 1 can be joined to any triple of distinct vertices by paths that overlap only in vertex 1; for example, $Q_1 = [1, 10, 4]$, $Q_2 = [1, 2, 6, 7]$, and $Q_3 = [1, 9, 8]$ join vertex 1, to vertices 4, 7, and 8, respectively.

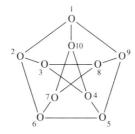

Figure 3.8. The Petersen graph

Suppose $U \subset V(G)$. If $o(U) \leq 3$, Theorem 3.20 guarantees the existence of a cycle containing every vertex of U. What if $o(U) = 4$ or 5? Given an arbitrary 4-vertex subset U of G, does there exist a cycle containing every vertex of U? In general, might the conclusion of Theorem 3.20 be strengthened to guarantee the existence of a cycle containing any prescribed set of $k + 1$ vertices? □

As expected, cycles have played a central role in our discussion of quantitative measures of connectivity. Let's finish the chapter with a somewhat lighter application of cycles.

Chap. 3   Connectivity

Suppose the edges of $G = K_4 - e$ are numbered in some fixed but arbitrary way. One possibility is pictured in Fig. 3.9. As shown in Fig. 3.10, G has three cycles, two of length 3 and one of length 4. If the highest numbered edge of a cycle is deleted, the result is a *broken cycle*. (See Fig. 3.11.)

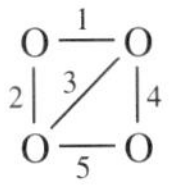

Figure 3.9

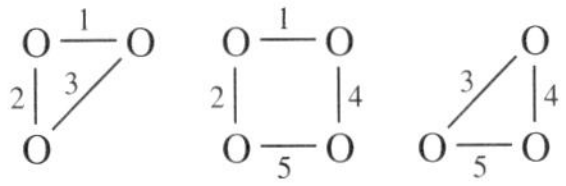

Figure 3.10

A *spanning tree* of G is a spanning subgraph that is a tree; a *spanning forest* is a spanning subgraph that is a forest —that is, a spanning subgraph that doesn't have any cycles.

3.22 Whitney's Broken Cycle Theorem[3]. *Let*

$$p(G, x) = x^n - c_1 x^{n-1} + c_2 x^{n-2} - \cdots + (-1)^{n-1} c_{n-1} x$$

be the chromatic polynomial of G. Then c_t is the number of t-edged spanning forests of G that do not contain a broken cycle.

3.23 Example.   With respect to the numbering of Fig. 3.9, $G = K_4 - e$ contains the three broken cycles in Fig. 3.11. Since it has five edges, G has five 1-edged spanning forests. Because no graph with a single edge contains a broken cycle, $c_1 = 5$.

There are $C(5, 2) = 10$ ways to choose a 2-edged spanning forest of G. The two that contain broken cycles have edge sets $\{1, 2\}$ and $\{3, 4\}$. (See Fig. 3.11.) Thus, $c_2 = 10 - 2 = 8$.

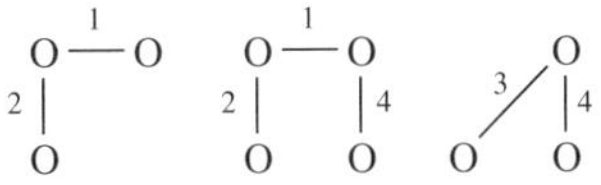

Figure 3.11

---

[3] A proof based on the Principle of Inclusion and Exclusion can be found in [H. Whitney, A logical expansion in mathematics, *Bull. Amer. Math. Soc.* **38** (1932), 572–579]. While not especially difficult, the techniques are beyond the scope of this book.

There are $C(5, 3) = 10$ ways to choose 3 edges from the 5 edges of G. From Fig. 3.10, the spanning subgraphs with edge sets $\{1, 2, 3\}$ and $\{3, 4, 5\}$ contain cycles. From Fig. 3.11, the spanning forests with edge sets $\{1, 2, 4\}$ and $\{1, 2, 5\}$ contain the broken cycle $\{1, 2\}$, and those with edge sets $\{1, 3, 4\}$ and $\{2, 3, 4\}$ contain the broken cycle $\{3, 4\}$. This leaves $\{1, 3, 5\}$, $\{1, 4, 5\}$, $\{2, 3, 5\}$, and $\{2, 4, 5\}$ as the edge sets of spanning forests that do not contain broken cycles; so, $c_3 = 4$. Thus, $p(G, x) = x^4 - 5x^3 + 8x^2 - 4x$. (Compare with Equation (13).) □

EXERCISES

1. Determine the edge connectivity, $\varepsilon(G)$, of $G =$

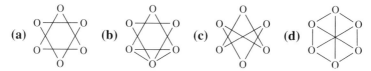

2. Determine the connectivity, $\kappa(G)$, of G in Exercise 1.

3. Exhibit a graph G with the following invariants:
   (a) $\delta(G) = 3$, $\varepsilon(G) = \kappa(G) = 0$.
   (b) $\delta(G) = 3$, $\varepsilon(G) = \kappa(G) = 1$.
   (c) $\delta(G) = \varepsilon(G) = 3$, $\kappa(G) = 1$.
   (d) $\delta(G) = \varepsilon(G) = 3$, $\kappa(G) = 2$.

4. Find the flaw in the following (much simpler) "proof" of Theorem 3.7: If $\varepsilon(G) = 0$, then $G = K_1$ or G is disconnected, in which cases $\kappa(G) = \varepsilon(G)$. Otherwise, suppose $\varepsilon(G) = t > 1$ and let $\{e_1, e_2, \ldots, e_t\}$ be a disconnecting set of edges of G. If $e_i = \{u_i, v_i\}$, $1 \leq i \leq t$, then $S = \{u_i : 1 \leq i \leq t\}$ is a vertex cut. Hence, $\varepsilon(G) = t \geq o(S) \geq \kappa(G)$.

5. Prove Corollary 3.4.

6. Suppose G is a connected graph on $n \geq 3$ vertices. Prove that $v \in V(G)$ is a cut-vertex of G if and only if there exist distinct vertices $u, w \in V(G)$, $u \neq v \neq w$, such that v is on every path in G from u to w.

7. Prove the following:
   (a) $\kappa(G)$ is an invariant. (b) $\varepsilon(G)$ is an invariant.

8. Suppose w_i, $1 \leq i \leq k$, are different vertices of a k-connected graph G. Let H be the graph obtained from G by adding a new vertex v^* and new edges v^*w_i, $1 \leq i \leq k$. Prove that H is k-connected.

9. Let G be a connected graph on n vertices, (at least) one of which is a cut-vertex. Prove that $(x - 1)^2$ is a factor of the chromatic polynomial $p(G, x)$.

10 Recall that a spanning tree is a spanning subgraph that is a tree.

  (a) Prove that G has a spanning tree if and only if it is connected. (Evidently, the number of different spanning trees in G is another quantitative measure of connectivity. We shall have more to say about the spanning tree number in Chapter 9.)

  (b) Suppose G is connected. Give a formula for the number of its spanning trees in terms of the numbers of spanning trees of its blocks. (*Hint:* Use the Fundamental Counting Principle.)

  (c) Suppose T is a spanning tree of G. Prove that no pendant vertex of T is a cut-vertex of G.

  (d) Let G be a connected graph on $n \geq 2$ vertices. Prove that at least two vertices of G are not cut-vertices.

11 Let u and v be nonadjacent vertices of G, where $\kappa(G) \geq 2$. Suppose P is a path in G from u to v. Prove or disprove that there exists a uv-path Q in G that is internally disjoint from P.

12 If B is a block of G, and $\varepsilon(G) \geq 2$, prove that $\kappa(B) \geq 2$.

13 Let G be a graph on $n \geq 3$ vertices. Suppose $\kappa(G) = 1$.

  (a) If B is a block of G, prove that B contains at least one cut-vertex of G.

  (b) Prove that G has (at least) two blocks, each of which contains exactly one cut-vertex of G.

  (c) Explain how parts *a* and *b* complete the proof of Theorem 3.13.

14 Prove that nonadjacent vertices u and v belong to different blocks of the connected graph G if and only if $\kappa_G(u, v) = 1$.

15 Prove that the edge set $E(G)$ is the disjoint union of $E(B)$ as B ranges over the blocks of G.

16 Suppose H is obtained from a 2-connected graph by subdividing one of its edges. (See Chapter 2, Exercise 30.) Prove that H is 2-connected.

17 Let G be a graph on $n \geq 3$ vertices with the property that given any 3-vertex subset U of $V(G)$, there exists a cycle of G that contains every vertex of U. Prove or disprove that G is 3-connected.

18 Suppose $G = (V, E)$ is a connected graph on $n \geq 3$ vertices. Let $F = \{\{u, v\} \in V^{(2)}: d(u, v) = 2\}$. Prove that $H = (V, E \cup F)$ is 2-connected.

19 Exhibit a graph G with two vertices u and v such that $\kappa(G) = 1$ and $\kappa_G(u, v) = 3$.

20 Suppose T is a tree on $n \geq 2$ vertices. Show that the blocks of T are all isomorphic to K_2.

21 Let G be a connected graph on $n \geq 2$ vertices, each of whose blocks is isomorphic to K_2. Prove that G is a tree.

22. Let u and v be nonadjacent vertices of the connected graph G. Suppose $S \subset V(G)$ is a minimum separating set for u and v, so that $\kappa_G(u, v) = o(S)$. Let H_u and H_v be the components of $G - S$ containing u and v, respectively. Prove that every $w \in S$ is adjacent in G to a vertex of H_u and to a vertex of H_v.

23. Let G be a 2-connected graph on $n \geq 3$ vertices. Prove that any two *edges* of G lie on a common cycle.

24. Suppose G is a 2-connected graph. Let g and c be the lengths of a shortest and a longest cycle of G, respectively. Then g is the *girth* of G, and c is its *circumference*.

    (a) Find the girth of the graph in Fig. 3.12.

    (b) Find the circumference of the graph in Fig. 3.12.

Figure 3.12

    (c) Prove that $g \leq 2d + 1$, where $d = \mathrm{diam}(G)$ is the diameter of G.

    (d) Prove or disprove that any two cycles of G of length c must have at least two vertices in common.

25. Let G be the graph in Fig. 3.12.

    (a) Find $\kappa(G)$.

    (b) Find $\psi_G(u, v)$.

    (c) Find $\psi_G(u, w)$.

26. Let $G = (V, E)$ be a connected graph. An *edge cut* (not to be confused with a cut-edge) of G is a set of edges of the form $\{uw \in E : u \in U$ and $w \in V \setminus U\}$, for some nonempty proper subset U of V.

    (a) Show that every edge cut is a disconnecting set of edges.

    (b) Show that every disconnecting set of edges contains an edge cut.

27. Let G be a graph with b blocks. Prove or disprove that the independence number $\alpha(G) \geq b$.

28. Let G be a graph. Prove the following:

    (a) $\chi(G) = \max \chi(B)$, as B ranges over the blocks of G.

    (b) $\omega(G) = \max \omega(B)$, as B ranges over the blocks of G.

Chap. 3 Connectivity

29 Let $G \neq K_n$ be a graph on n vertices. Suppose k is a positive integer satisfying $k \leq 2\delta(G) - (n-2)$. Prove that $\kappa(G) \geq k$.

30 (Whitney) Let G be a graph on $n \geq 3$ vertices. Prove that G is k-connected if and only if for every pair of distinct vertices u and v of G, there exist (at least) k internally disjoint paths in G from u to v.

31 Find the (common) chromatic polynomial for the three graphs in Fig. 3.3.

32 Consider the sequence $\pi = (3, 3, 2, 2, 2)$.

(a) Find nonisomorphic graphs G and H (on $n = 5$ vertices) such that $d(G) = \pi = d(H)$.

(b) Show that one of G and H (see part (a)) is bipartite and the other is not.

(c) Show that both G and H (see part (a)) are 2-connected.

33 Let G and H be connected graphs. Suppose there is a one-to-one correspondence between the blocks of G and the blocks of H in which corresponding blocks are isomorphic. Prove the following:

(a) G and H have the same number of edges.

(b) G and H have the same number of vertices.

(c) $p(G, x) = p(H, x)$.

34 Let G be a graph with b blocks. Suppose that, taken separately as in Figs. 3.1 and 3.2, the b blocks of G have a total of n vertices altogether. Prove that $o(V(G)) \geq n - b + 1$, with equality if and only if G is connected.

35 Of the 34 graphs on 5 vertices, 21 are connected and 10 are blocks (2-connected). Illustrate the nonisomorphic 3-connected graphs on 5 vertices.

36 Let $G = Q_k$, the cube graph on $n = 2^k$ vertices defined in Chapter 1, Exercise 33.

(a) Find $\kappa(G)$. (b) Find $\varepsilon(G)$.

37 Let G be the Petersen graph, illustrated in Fig. 3.8.

(a) Show that $\kappa(G) > 2$.

(b) Find the girth of G. (See Exercise 24.)

(c) Find the circumference of G. (See Exercise 24.)

38 Show that the conclusion of Theorem 3.20 *cannot* be strengthened to guarantee the existence of a cycle containing any prescribed set of 3 vertices of an arbitrary 2-connected graph.

39 Give an alternative proof of the $k = 2$ case of Theorem 3.20 based on Menger's Theorem.

40 Let G be a graph with n vertices and m edges. Use Theorem 3.22 to prove that $p(G, x) = x^n - mx^{n-1} + \cdots$.

41 If the edges of $G = K_4 - e$ are numbered as in Fig. 3.9, then the graph H obtained from G by deleting edge number 5 is isomorphic to the overlap of K_3 with K_2 in K_1.

  (a) Illustrate the single broken cycle of H.

  (b) How many 2-edged spanning subgraphs does H have?

  (c) How many of the 2-edged spanning subgraphs of H contain its broken cycle?

  (d) Illustrate the 3-edged spanning forests of H that do not contain its broken cycle.

  (e) Use Whitney's Broken Cycle Theorem to compute $p(H, x)$.

42 Use Whitney's Broken Cycle Theorem to give a new proof that $p(T, x) = x(x - 1)^{n-1}$ for any tree T on n vertices. (*Hint:* By the Binomial Theorem, the coefficient of x^{n-1-k} in $(x - 1)^{n-1}$ is $(-1)^k C(n - 1, k)$.)

43 Let $G = K_{2,3}$.

  (a) Show that the chromatic polynomial of G is $p(G, x) = x(x - 1)(x^3 - 5x^2 + 10x - 7)$.

  (b) Show that $p(G, x) = 0$ has a solution in the interval $(1,2)$.

44 Let $G = K_{3,4}$.

  (a) Show that G is 3-connected.

  (b) Given that the chromatic polynomial of G is $p(G, x) = x(x - 1)(x^5 - 11x^4 + 55x^3 - 147x^2 + 204x - 115)$, show that $p(G, x) = 0$ has a solution $x \doteq 1.781$.[4] (*Hint:* Use Newton's Method.)

---

[4] Let G be a graph. It was shown by W. T. Tutte [Chromials, in *Hypergraph Seminar* (C. Berge and D. K. Ray-Chaudhuri, eds.), *Lecture Notes in Math.* **411** (1974), 243–266] that $p(G, x) \neq 0$ for all $x \in (0, 1)$, and by B. Jackson [A zero-free interval for chromatic polynomials of graphs, *Comb. Prob. Comp.* **2** (1993), 325–336] that $p(G, x) \neq 0$ for all $x \in (1, 32/27)$. Jackson conjectured that if G is 3-connected and not bipartite, then $p(G, x) \neq 0$ for all $x \in (1, 2)$. That G must be nonbipartite in Jackson's conjecture is demonstrated by this exercise. More generally, D. R. Woodall [Zeros of chromatic polynomials, in *Surveys in Combinatorics* (P. J. Cameron, ed.), Academic Press, 1977, pp 199–223] showed that if $r, s \geq 2$ have different parity, then $p(K_{r,s}, x) = 0$ *always* has a solution $x \in (1, 2)$.

4

Planar Graphs

When graphs are pictured by dots and arcs, it is typically the case that some nonadjacent edges are illustrated by arcs that *cross* — that is, intersect at a point that does not represent a vertex. It is natural to wonder when and how such potentially misleading representations can be avoided. When is it possible to (re)draw a graph in such a way that untidy edge crossings do not occur? Evidently (Fig. 4.1), it is possible to draw K_4 with no edge crossings, but what about K_5?

Figure 4.1

Given enough room, it is always possible to draw a graph, any graph, in such a way that no two edges cross. Suppose G has n vertices and m edges. Represent the vertices by n distinct points along the z-axis in real, three-dimensional Euclidean space. Then draw an edge of G in each of m different planes that intersect (only) in the z-axis.

What about two-dimensional space? Which graphs can be drawn in the Euclidean plane with no edge crossings? This is a much more interesting problem, not because its solution has any important applications but because the search for a solution has led to some good mathematics.

4.1 Definition. A graph is said to be *planar* if it can be illustrated in the plane so that arcs representing edges meet only at points representing vertices.

Informally, G is planar if it can be drawn in the plane without any edge crossings. Such a drawing is referred to as a *plane graph*, or as an *embedding* of G in the plane.

Any discussion of plane graphs leads eventually to the concept of a "region." If the locus of points comprising an embedding of G is deleted from two-dimensional

Euclidean space, the connected components of what remains are *regions* of G. Of these regions, one is unbounded and the rest are bounded.

It is natural to wonder how the number of regions can vary over different embeddings of the same planar graph G. Can one embedding of G yield 6 regions and another 8? If so, must there always be an embedding with a total of 7 regions? The definitive answer to such questions was discovered by Euler.[1]

4.2 Euler's Formula (for plane graphs). *Let G be a plane graph with n vertices, m edges, c components, and $r = r(G)$ regions. Then*

$$r = m - n + c + 1. \qquad (22)$$

Proof. The proof is by induction on m. If $m = 0$, then $G = K_n^c$ is a graph consisting of $c = n$ components, each of which is an isolated vertex. Because K_n^c has just one (unbounded) region, Equation (22) reduces to the identity $1 = 1$.

Assume the result is true for every plane graph having $k \geq 0$ edges. Let G be a plane graph with $k + 1$ edges, and suppose e is one of them. If the same region lies on both sides of e, then e is a bridge, in which case $G - e$ is a plane graph with the same numbers of vertices and regions as G, but with one fewer edge and one more component. Applying the induction hypothesis to $G - e$ produces

$$r(G) = r(G - e)$$
$$= (m - 1) - n + (c + 1) + 1$$
$$= m - n + c + 1.$$

If e is part of the boundary separating (two) different regions, then one (at least) of the regions is bounded, so e lies on some cycle of G. In this case, $G - e$ is a plane graph having the same numbers of vertices and components as G, but one fewer edge and one fewer region. Applying the induction hypothesis to $G - e$, we obtain $r(G) - 1 = r(G - e) = (m - 1) - n + c + 1$ which is equivalent to Equation (22). ∎

In the special case that G is a *connected* plane graph, Equation (22) can be expressed in the form

$$r + n = m + 2. \qquad (23)$$

Following in the footsteps of Ptolemy of Alexandria, the Flemish cartographer Gerhard Mercator (1512–1594) popularized the technique of mapmaking in which meridians (lines of longitude) are drawn parallel to each other and perpendicular to the lines of latitudes. Details aside, Mercator "projected" the surface of a sphere into the surface of a plane. The same sort of thing can be done with any convex polyhedron. Consider, for example, Fig. 4.2, in which the surface of a regular icosahedron has been projected onto a plane. Just as the

---

[1] A contemporary of Benjamin Franklin (1706–1790), Leonhard Euler (1707–1783) was the most prolific mathematician of all time.

Chap. 4 Planar Graphs

typical mercator projection of Earth makes Greenland appear comparable in size to South America, the congruent equilateral triangular faces of the icosahedron have been badly distorted. In fact, one of the 20 faces becomes the unbounded "exterior" region of the resulting plane graph!

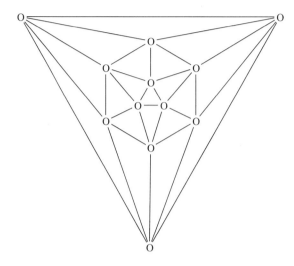

Figure 4.2

In any such projection, there are natural one-to-one correspondences between the vertices of the polyhedron and the vertices of the graph, between the edges of the polyhedron and the edges of the graph, and between the faces of the polyhedron and the regions of the graph. Consequently, Equation (23) establishes a relationship between the numbers of faces (F), vertices (V), and edges (E) of any convex polyhedron, namely,

$$F + V = E + 2. \tag{24}$$

4.3 Corollary. *If G is a planar graph with $n \geq 3$ vertices and m edges, then*

$$m \leq 3n - 6. \tag{25}$$

Proof. Let G be a plane graph. It may happen that some nonadjacent pair of its vertices can be joined by a new edge, drawn so that it does not cross any of the existing edges of G. (Remember that arcs representing edges need not be straight line segments.) A *triangulation* of G is a plane graph H that contains G as a spanning subgraph, and to which no new edge can be added without crossing some existing edge. In Fig. 4.2, $H = G$ is already *triangulated*. In this particular case, $n = 12, m = (12 \times 5)/2 = 30 = 3n - 6$.

In general, suppose H is a triangulation of G having n vertices and $m + k$ edges. We will complete the proof by showing that $m + k = 3n - 6$. Evidently, H

is connected; otherwise, it would be possible to add a bridge between some pair of components (without crossing an existing edge of H). Moreover, the length of the boundary cycle (*perimeter*) of each region of H, including the unbounded region, must be three; otherwise, a *chord* of the cycle could be added to $E(H)$ without crossing any existing edges. If H has a total of r regions, then, by counting the edges around the perimeter of each of them and summing over the regions, we obtain a total of $3r$ edges. In this total, each edge has been counted exactly twice; that is, $3r = 2(m + k)$. Solving for r and substituting the result into Equation (23) yields $2(m + k)/3 + n = (m + k) + 2$; that is, $m + k = 3n - 6$. ∎

4.4 Example. The complete graph, K_5, has $n = 5$ vertices and $m = 10$ edges. If K_5 is planar, then, from Corollary 4.3, $10 \leq 3 \times 5 - 6$. So, K_5 is *not* planar. □

4.5 Corollary. *Suppose G is a bipartite planar graph with m edges and $n \geq 3$ vertices. Then*

$$m \leq 2n - 4. \qquad (26)$$

Proof. Suppose G is a bipartite plane graph. By Theorem 2.23, G has no odd cycles. It may happen that some nonadjacent pair of vertices of G can be joined by a new edge that crosses none of the existing edges and does not create an odd cycle. Suppose a maximum number k of such edges have been added to G. Call the resulting bipartite plane graph H. This time, each of the r regions of H has a perimeter of length four and it follows that $2(m + k) = 4r$. Substituting $r = (m + k)/2$ into Equation (23) yields $(m + k)/2 + n = (m + k) + 2$; that is, $2n - 4 = m + k \geq m$. ∎

4.6 Example. The complete bipartite graph, $K_{3,3}$, has $n = 6$ vertices and $m = 9$ edges. If $K_{3,3}$ is planar, then, from Corollary 4.5, $9 \leq 2 \times 6 - 4$. So, $K_{3,3}$ is not planar. □

Suppose G is a plane graph. Then any graph obtained from G by removing some of its vertices and/or edges must also be a plane graph. Said another way, if H is not planar, then any graph containing a subgraph isomorphic to H cannot be planar. So far, we have identified two such *forbidden* subgraphs, namely, K_5 and $K_{3,3}$. Given a forbidden subgraph H, we can construct others by "subdividing" its edges. Informally, to subdivide $e \in E(H)$ means to put a new vertex of degree 2 in the middle of e.

4.7 Definition. Let $H = (W, F)$ be a graph. Suppose that $e = uw \in F$ and $v \notin W$. Let $G = (V, E)$, where $V = W \cup \{v\}$ and $E = (F \setminus \{uw\}) \cup \{uv, vw\}$ is the set obtained from F by replacing $e = uw$ with two new edges, uv and vw. Then G is a graph obtained from H by *subdividing e*.

Suppose H has n vertices and m edges. If G is obtained from H by subdividing one of its edges, then G has $n + 1$ vertices and $m + 1$ edges. Moreover, any plane drawing of H can be modified to obtain an embedding of G and vice versa; H is planar if and only if G is planar. The same thing could be said no matter how many edges of H were subdivided. A *subdivision* of H is any graph that can be obtained from H by subdividing edges. Thus, e.g., P_n is a subdivision of P_2 for all $n \geq 3$ and C_n is a subdivision of C_3 for all $n \geq 4$. While the graph G_1 in Fig. 4.3 is a subdivision of H, the graph G_2 is not.

Figure 4.3

4.8 Definition. Two graphs are *homeomorphic* if they have isomorphic subdivisions.

Informally, homeomorphic graphs are isomorphic "to within vertices of degree two." Evidently, any graph is homeomorphic to any of its subdivisions. Finally, and this is the point of the subdivision idea, if G has a subgraph homeomorphic to K_5 or to $K_{3,3}$, then G cannot be planar. In 1930, Kasimir Kuratowski proved the converse.[2]

4.9 Kuratowski's Theorem. *Let G be a graph. If G does not have a subgraph homeomorphic to K_5 or to $K_{3,3}$, then G is planar.*

Omitting the long, technical proof of Kuratowski's Theorem, we return to another of the many consequences of Euler's Formula.

4.10 Lemma. *If G is a planar graph with n vertices and m edges, then $\delta(G) \leq 5$.*

Proof. If $\delta(G) \geq 6$, then $2m = \sum d(v) \geq 6n$, contradicting Corollary 4.3. ∎

4.11 Five-Color Theorem. *Any planar graph can be colored properly with five colors; that is, if G is planar, then $\chi(G) \leq 5$.*

Proof. Because *any* graph on $n \leq 5$ vertices can be colored properly with five colors, a proof by induction on the number of vertices is off to a good start. So, assume $\chi(H) \leq 5$ for every planar graph H on $k \geq 5$ vertices, and let G be a plane graph on $n = k + 1$ vertices. By Lemma 4.10, G has a vertex w of degree at most five. If $H = G - w$, then, by the induction hypothesis, $\chi(H) \leq 5$.

---

[2] K. Kuratowski, Sur le problème des courbes gauches en topologie, *Fund. Math.* **15** (1930), 271–283.

If $\chi(H) < 5$, then any proper four-coloring of H can be extended to a proper five-coloring of G by giving w the fifth color. If $d_G(w) < 5$, then any proper five-coloring of H can be extended to G by giving w a color not assigned to any of its neighbors. Hence, we proceed under the assumption that $\chi(H) = 5 = d_G(w)$.

Vertex w and its five neighbors in the plane graph G are illustrated in Fig. 4.4. If, in some proper five-coloring of H, two neighbors of w happened to be colored the same, then, once again, the proper coloring of H can be extended to G. This brings us to the hard case. Given a proper five-coloring of H, assume that vertex v_i is colored c_i, $1 \leq i \leq 5$, and that these five colors are all different.

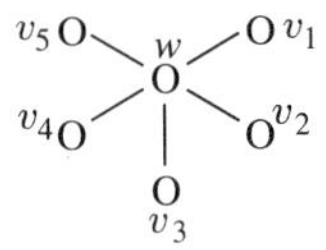

Figure 4.4

Suppose there is a path P in H from v_2 to v_4, the vertices of which are colored alternately c_2 and c_4. Then $C = \langle P, w \rangle$ is a cycle in the plane graph G. It may be that v_3 is inside C, as shown in Fig. 4.5. The alternative is that v_3 is outside C while v_1 and v_5 are inside. Either way, there could not exist a path in H from v_1 to v_3, the vertices of which are colored alternately c_1 and c_3. (A path in H from v_1 to v_3 is a path in G, so it cannot cross an edge of C. Because the colors are wrong, it cannot pass through a vertex of C.) Evidently, H cannot contain both an alternating $c_1 - c_3$ path from v_1 to v_3 and an alternating $c_2 - c_4$ path from v_2 to v_4. Without loss of generality, we may assume that H does not contain an alternating $c_1 - c_3$ path from v_1 to v_3.

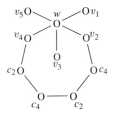

Figure 4.5

It might happen that no vertex of H is both adjacent to v_1 and colored c_3. If so, we could change the color of v_1 from c_1 to c_3 and free up color c_1 for w. The rest of the proof is a generalization of this idea: Let X be the set of those vertices x of H such that there is an alternating $c_1 - c_3$ path in H from v_1 to x. It is our working assumption that $v_3 \notin X$. Now, the crucial observation about X is that if $v \in V(H)$ is colored either c_1 or c_3, and if v is adjacent to a vertex of X, then $v \in X$. Put another way, if $v \notin X$, but v is adjacent to a vertex of X, then v is colored neither c_1 nor c_3. Consequently, if we switch the colors of the vertices in X, the result is a new, proper five-coloring of H, one in which both v_1 and v_3 are colored c_3. This new coloring may be extended to a proper five-coloring of G by assigning c_1 to w. ∎

In the last "hard" case of the proof, vertex v_5 appears to be superfluous. It seems that it ought to be possible, somehow, to eliminate v_5 and obtain a proof of the following.

4.12 Four-Color Theorem. *Any planar graph can be colored properly with four colors; that is, if G is planar, then $\chi(G) \leq 4$.*

In 1850, Francis Guthrie noticed that four colors suffice to distinguish the counties on a map of England and wondered whether four colors are enough to color the regions of any plane graph in such a way that regions whose perimeters share an edge are colored differently. Guthrie's younger brother, Frederick, shared the problem with his teacher, Augustus de Morgan. In 1852, de Morgan passed the problem on to William Rowan Hamilton. Eventually, in 1878, Arthur Cayley formally presented the problem at a meeting of the London Mathematical Society. The next year, British barrister Alfred Kempe published a "proof" of the Four-Color Theorem in the *American Journal of Mathematics*. Several years later, Percy Heawood found an error in Kempe's proof. A century later, Kenneth Appel and Wolfgang Haken reduced the problem to a consideration of 1476 "configurations." Using hundreds of hours of computer time, they managed to show that four colors suffice for each of them, thus proving the Four-Color Theorem.[3] More recent work has reduced the number of configurations that have to be checked even as faster computers have reduced the time to check each one. Still, as of this writing, no one has found a proof of the Four-Color Theorem that is both correct and computer-free.[4]

The connection between the original four-color problem and proper (vertex) colorings of graphs is through the notion of a geometric *dual*, an idea going back to Euclid. Let G be a plane graph. Denote the set of its regions by $F(G) = \{f_1, f_2, \ldots, f_r\}$. The vertex set of the geometric dual is $V(G^d) = F(G)$. The edges of G^d are in one-to-one correspondence with the edges of G. Given $e \in E(G)$, suppose f_i and f_j are the regions of G on either side of e. Then the edge of G^d corresponding to e is $f_i f_j$.

4.13 Example. It is usually convenient to draw G^d right on top of G, with each edge of G^d crossing the edge of G to which it corresponds. The situation for $G = K_3$ is illustrated in Fig. 4.6. The bad news is that G^d can be a multigraph—or

---

[3] See K. Appel and W. Haken, *Every Planar Map Is Four Colorable*, Contemporary Mathematics, Vol. 98, Amer. Math. Soc., Providence, R.I., 1989.

[4] This situation is frequently cited in philosophical discussions about the nature of mathematical proof.

worse. As illustrated in Fig. 4.7, G^d can be a *pseudograph*, an object that contains *loops* as well as multiple edges. (A *loop* is an "edge" from a vertex to itself.) Indeed, it is easily seen that loops in G^d correspond to bridges in G. Multiple edges occur when the perimeters of two regions of G share more than one edge. □

Figure 4.6

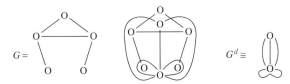

Figure 4.7

4.14 Example. The graphs G_1 and G_2 illustrated in Fig. 4.8 are isomorphic; that is, they are two different embeddings of the same planar graph G. It may be somewhat surprising, therefore, to discover that $G_1^d \not\cong G_2^d$. Put another way, the graphs illustrated in Fig. 4.8 are isomorphic, but their dual multigraphs are not! (While the confirmation of this fact is left to the exercises, why not take a minute to do it now?)[5] □

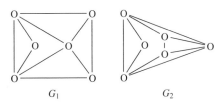

Figure 4.8. Isomorphic plane graphs with nonisomorphic duals.

Despite the complications illustrated in these examples, every dual pseudo/multigraph G^d has a unique underlying graph G_d obtained from G^d by removing

---

[5] It was shown by H. Whitney [Congruent graphs and the connectivity of graphs, *Amer. J. Math.* **54** (1932), 150–168] that this sort of thing cannot happen with 3-connected plane graphs.

Chap. 4 Planar Graphs

loops and/or duplicate edges. (The underlying graph obtained from G^d in either Fig. 4.6 or 4.7 is $G_d = K_2$.)

If the vertex incident with a loop is adjacent to itself, no pseudograph could be colored properly. Thus, it is customary to ignore loops when coloring vertices. It follows from this convention that $\chi(G^d) = \chi(G_d)$. Therefore, coloring regions of G is the same as coloring vertices of G_d. Because G_d is planar, the Four (Map)-Color Theorem is a consequence of Theorem 4.12.

What about embedding graphs in other surfaces? It is not hard to see that a graph can be embedded in a plane if and only if it can be embedded in (the surface of) a sphere.[6] A formal proof of this fact could be constructed along the following lines: Imagine a sphere S placed on a plane Π in such a way that the south pole of S corresponds to the origin, O, of Π. Let $P \neq O$ be a point of Π. If L is the line segment whose endpoints are P and the north pole of S, then the interior of L intersects S in a unique point, call it $f(P)$. Define $f(O)$ to be the south pole of S. Then f is a one-to-one function from Π onto the *punctured sphere* S' obtained from S by deleting its north pole. Moreover, any graph G embedded in Π is mapped by f to an embedding of G in S'.

Conversely, any graph G that can be embedded in the sphere can be embedded in such a way that the north pole is in the interior of some region of G. For any such embedding, $f^{-1}(G)$ is a plane graph (whose unbounded region corresponds to the region of G that contains the north pole). Thus, as far as embedding graphs is concerned, planes and punctured spheres are equivalent surfaces. In particular, neither $K_{3,3}$ nor K_5 can be embedded in a sphere.

Figure 4.9(a) shows K_5 embedded in a torus. It is, of course, the hole that makes the difference. Moreover, just as a punctured sphere is "topologically equivalent" to a plane, a punctured torus is topologically equivalent to a plane with an overpass. Figure 4.9(b) illustrates K_5 embedded in a plane with an overpass.

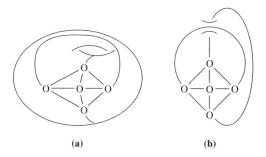

(a)   (b)

Figure 4.9

A plane with a finite number of overpasses is an *orientable surface*. The number of overpasses is the *genus* of the surface. Pretty clearly, any graph can

---

[6] That is, drawn on the surface of a sphere in such a way that no two of its edges cross. Indeed, this observation is implicit in the equivalence of Equations (23) and (24).

be embedded in an orientable surface of sufficiently large genus: Let G be a fixed but arbitrary graph. Draw a picture of G in the plane. Adjust the picture if necessary so that at most two edges cross in any point (not representing a vertex). Then erect an overpass at every point where two edges of G cross.

4.15 Definition. Let G be a graph. The *genus* of G, denoted $\gamma(G)$, is the minimum number of overpasses that must be added to the plane so that G can be embedded in the resulting surface.

If G_1 and G_2 are isomorphic graphs, they share the same pictures. Thus, it is an immediate consequence of Definition 4.15 that γ is an invariant. It is the minimum genus of the orientable surfaces on which G can be drawn with no edge crossings. In particular, G is a planar graph if and only if $\gamma(G) = 0$. Because K_5 is not planar, Fig. 4.9(b) proves that $\gamma(K_5) = 1$. Just as the phrase "plane graph" refers to a particular embedding of a planar graph, a "*bagel* graph" is a particular toroidal embedding of a graph of genus 1.

Consider the toroidal embeddings of $G = K_4$ illustrated in Fig. 4.10. When G is deleted from the right-hand embedding, the result is a disconnected surface having four components, also known as *regions*. On the other hand, deleting G from the left-hand embedding leaves not four regions, but three! It seems the number of regions that result when a graph is embedded in a surface of positive genus depends on the embedding. It turns out, however, that this phenomenon cannot occur when $\gamma(G)$ coincides with the genus of the surface. *If G is embedded in an orientable surface S of genus $\gamma(G)$, then the number of connected components of $S \backslash G$ is independent of the embedding.* Said another way, if G is drawn, with no edge crossings, on a plane having $\gamma(G)$ overpasses, the resulting number of regions is a function (only) of G. This independence from the embedding is a consequence of the following analog of Theorem 4.2.

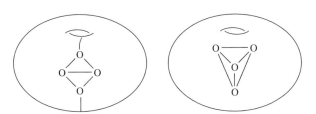

Figure 4.10

4.16 Euler's Formula (for connected graphs of arbitrary genus). *Suppose G is a connected graph with n vertices and m edges. If G is embedded in an orientable surface S of genus $\gamma(G)$, then the resulting number of regions is*

$$r(G) = m - n + 2 - 2\gamma(G). \qquad (27)$$

When G is planar, $\gamma(G) = 0$, and Equation (27) evolves into Equation (23).

Chap. 4 Planar Graphs   73

4.17 Example. Figure 4.9 exhibits the bagel graph $G = K_5$. From equation (27), $r(G) = 10 - 5 + 2 - 2 = 5$. To confirm this result, imagine the vertices in Fig. 4.9(b) to be fence posts (viewed from above) and the edges to be fences on the orientable surface, T. It is easy to identify four triangular regions which, taken together, might be described as a *compound*. As shown in Fig. 4.11, this compound is inside the cycle $C = \langle v, y, u, x \rangle$. We claim that the points of $T \backslash G$ outside C comprise a single region. To test this hypothesis, we are going to take a walk. Starting at a point outside the compound, midway between vertex (fence post) v and vertex y, walk along edge (fence) vy to vertex y. Keeping the fence on the left, continue long edge yx to vertex x.

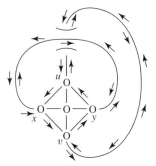

Figure 4.11

The idea is to continue walking, within reach of the fence to our left, until we return to our starting point. During this walk, any point on our right, that can be reached without climbing a fence, is outside the compound. By the end of the walk it should be clear that any two points outside the compound can be joined by an arc that roughly parallels our walk and crosses none of the fences — that is, that the points of $T \backslash G$ not inside the compound comprise a single region. This confirms that the embedding creates exactly five regions, one outside and four inside the compound.

While this perambulation on the orientable surface may have clarified one issue, it has raised some others. Taken in order, the vertices of G that we encountered on our recent walk were y, x, v, u, x, y, u, v. It seems we walked along every edge of the cycle C exactly once, but along xy and uv twice! Moreover, while we returned to the starting point without retracing any steps, our route was not along any cycle of G! Among the problems of a rancher grazing cattle on the surface is what to do about fences xy and uv. Among the problems for the graph theorist is how to describe the boundary of the outside region. (Such problems do not arise for graphs embedded in the plane!) □

4.18 Definition. A *walk* in G of *length* k, from v to w, is a sequence $v = u_0, u_1, \ldots, u_k = w$ of vertices of G such that $u_{i-1}u_i \in E(G)$, $1 \leq i \leq k$. If $w = v$, the walk is said to be *closed*.

Evidently, a path is a walk in which the vertices are all different, and a cycle is a closed walk all of whose vertices are distinct.

Suppose R is a region created by the embedding of some graph G in a surface of genus $\gamma > 0$. Then the perimeter (boundary) of R is a closed walk in G. If R lies on both sides of some $e \in E(G)$, then e occurs exactly twice along its perimeter. If no such e exists, then the perimeter of R is a cycle of G.

4.19 Theorem. *Let G be a connected graph with $n \geq 3$ vertices and m edges. Then*

$$\gamma(G) \geq \frac{m}{6} - \frac{n}{2} + 1. \tag{28}$$

Proof. If $n = 3$, then $\gamma(G) = 0$, and the result follows from the fact that $m \leq 3$. Otherwise, let r be the number of regions created by an embedding of G in an orientable surface of genus $\gamma(G)$. Summing the numbers of edges around the perimeter of each region (multiplicities included) we obtain a total of $2m$ because, in this sum, each edge is counted twice. On the other hand, because the perimeter of a region has at least three edges, the total can be no less than $3r$. So, $2m \geq 3r$. When this inequality is used to eliminate $r = r(G)$ from Equation (27), the result is equivalent to (28). ∎

When $G = K_n$, Inequality (28) becomes

$$\begin{aligned}
\gamma(K_n) &\geq \frac{C(n,2)}{6} - \frac{n}{2} + 1 \\
&= \frac{n(n-1)}{12} - \frac{6n}{12} + \frac{12}{12} \\
&= \frac{n^2 - 7n + 12}{12}.
\end{aligned} \tag{29}$$

Rewrite (29) as

$$12\gamma_n - 12 \geq n^2 - 7n,$$

where $\gamma_n = \gamma(K_n)$. Completing the square yields

$$12\gamma_n + \frac{1}{4} \geq \left(n - \frac{7}{2}\right)^2.$$

Taking square roots and rearranging terms, we obtain

$$\frac{7 + \sqrt{48\gamma_n + 1}}{2} \geq n$$

$$= \chi(K_n), \tag{30}$$

Chap. 4 Planar Graphs

a relation between the genus and chromatic numbers of K_n. Percy Heawood[7], was able to extend Inequality (30) so as to encompass all *nonplanar* graphs.

4.20 Heawood's Theorem. *If $\gamma(G) > 0$, then*

$$\left\lfloor \frac{7 + \sqrt{48\gamma(G) + 1}}{2} \right\rfloor \geq \chi(G). \tag{31}$$

Setting $\gamma(G) = 0$ in (31) yields $4 \geq \chi(G)$, precisely the Four-Color Theorem. Imagine how frustrating it must have been for Heawood to be able to prove Inequality (31) for all graphs of positive genus, yet not to be able to prove it when $\gamma(G) = 0$.

When $\gamma(G) = 1$, (31) becomes $7 \geq \chi(G)$. Can that be a tight upper bound? Is there a bagel graph whose vertices cannot be colored properly with six colors? More generally, denote a plane with p overpasses by Π_p. (Then Π_p is an orientable surface of genus $\gamma = p$.) Let $\chi(\Pi_p) = \max \chi(G)$, where the maximum is over all graphs G that can be embedded in Π_p. By the Four-Color Theorem, $\chi(\Pi_0) = 4$. What is the value of $\chi(\Pi_p)$ in general? The answers to these questions emerge in an interesting way. Instead of completing the square in Inequality (29), factor the numerator so as to obtain

$$\gamma(K_n) \geq \frac{(n-3)(n-4)}{12}. \tag{32}$$

Because it must be an integer,

$$\gamma(K_n) \geq \left\lceil \frac{(n-3)(n-4)}{12} \right\rceil. \tag{33}$$

Heawood believed Inequality (33) to be an identity, a result that was finally proved by Ringel and Youngs in 1968.[8]

4.21 Theorem. *The genus* $\gamma(K_n) = \left\lceil \dfrac{(n-3)(n-4)}{12} \right\rceil$, $n \geq 3$.

It follows from Theorem 4.21 that $\gamma(K_7) = 1$; that is, K_7 has a toroidal embedding. Since $\chi(K_7) = 7$, this embedding is a bagel graph that cannot be colored properly with only 6 colors. This proves that Inequality (31) is tight for graphs of genus 1; that is, $\chi(\Pi_1) = 7$. More generally,

$$\chi(\Pi_p) = \left\lfloor \frac{7 + \sqrt{48p + 1}}{2} \right\rfloor$$

for all $p \geq 0$.

---

[7] P. J. Heawood, Map colour theorems. *Quart. J. Pure Appl. Math.* **24** (1890), 332–338.
[8] G. Ringel and J. W. Youngs, Solution of the Heawood map-coloring problem, *Proc. Nat. Acad. Sci.* **60** (1968), 438–445.

What about Kuratowski's Theorem? Does it have an analog, say, for bagel graphs? It is known that, for every nonnegative integer p, there exists a finite set of forbidden graphs $F(p)$ such that G can be embedded in an orientable surface of genus p if and only if it does not have a subgraph homeomorphic to an element of $F(p)$. But, this result is existential. No one has found an exact analog of Kuratowski's Theorem for any $p > 0$.

EXERCISES

1 Prove that every tree is a planar graph.

2 Use Equation (23) and Exercise 1 to prove that every tree on n vertices has $n - 1$ edges. (Compare with Chapter 2, Exercise 31.)

3 Let G be a planar graph on $n \geq 4$ vertices. Prove that G has at least four vertices of degrees at most five.

4 Prove that G is planar if and only if each of its blocks is planar.

5 In 1936, K. Wagner proved that every planar graph has an embedding in which each edge is represented by a *straight* line segment.[9] Draw such an embedding of

   (a) $K_5 - e$.   (b) $K_{3,3} - e$.   (c) C_6^c.

6 Prove that every triangulated plane graph on $n \geq 3$ vertices is 2-connected.

7 Prove or disprove the converse of the Four-Color Theorem.

8 Suppose $e = uv \in E(G)$. Recall that G/e is obtained from G by contracting e — that is, by deleting edge e, coalescing vertices u and v into a single vertex, and deleting any multiple edges that may have arisen in the process. Say that G is *contractible* to H if H can be obtained from G by a sequence of contractions. It was proved by Wagner[10] that G is planar if and only if it does not have a subgraph that is contractible to K_5 or to $K_{3,3}$. Prove that the Petersen graph is not planar

   (a) using Wagner's criterion.

   (b) using Kuratowski's criterion.

9 In 1985, chemists synthesized the first fullerene,[11] a molecule, C_{60} (not to be confused with the cycle of length 60), consisting of 60 carbon atoms — and nothing else. The existence of this third form of carbon (the first two being

---

[9] K. Wagner, Bemerkungen zum Vierfarbenproblem, *Jber. Deutsch. Math.-Verein.* **46** (1936), 26–32. The result was discovered independently by I. Fáry, On the straight line representations of planar graphs, *Acta Sci. Math.* **11** (1948), 229–233.

[10] K. Wagner, Über eine Eigenschaft der ebenen Komplexe, *Math. Ann.* **114** (1937), 570–590.

[11] Robert Curl, Harold Kroto, and Richard Smalley received the 1996 Nobel Prize in Chemistry for this work. They were assisted by graduate students J. R. Heath and S. C. O'Brien.

graphite and diamond) had been predicted by R. Buckminster Fuller. Less expected were C_{70}, C_{76}, C_{84}, C_{90}, and others, all of which had been produced by 1992. Molecular C_{60} comes in a recognizable form. Known to mathematicians as a *truncated icosahedron*, it is one of the 13 semiregular polyhedra classified by Archimedes in the third century BC. (See Fig. 4.12.) While every fullerene takes the shape of a convex polyhedron each of whose faces is a pentagon or a hexagon, none of the higher fullerenes is semiregular.

Figure 4.12

(a) Prove that the number of pentagonal faces in each (and every) fullerene is exactly 12.

(b) Prove that every fullerene molecule contains an even number of carbon atoms.

(c) Compute the number of hexagonal faces in molecular C_n.

(d) Compute the number of hexagons on a soccer ball.

10 Let G be a plane graph.

(a) Suppose $\varepsilon(G) \geq 2$ and $\delta(G) \geq 3$. Prove that G^d is a graph (i.e., $G^d = G_d$).

(b) Suppose G is 3-connected. Prove that G^d is a graph.

11 Graph G is *coplanar* if both G and G^c are planar.[12]

(a) Suppose G is a graph on $n \leq 5$ vertices that is not coplanar. Prove that $G = K_5$ or $G = K_5^c$.

(b) Exhibit a coplanar graph G on 6 vertices such that both G and its complement are connected.

(c) Let G be a coplanar graph on n vertices. Prove that $n < 11$.

---

[12] D. Cvetković, A. Jovanović, Z. Radosavljević, and S. Simić, Coplanar graphs, *Publikacije Elek. Fak. Univ. Belgrade* **2** (1991), 67–81.

12 Let G_1 and G_2 be the plane graphs illustrated in Fig. 4.8. Prove the following:
   (a) G_1 and G_2 are isomorphic.
   (b) G_1 and G_2 have nonisomorphic dual multigraphs.
   (c) The graphs underlying the duals of G_1 and G_2 are isomorphic.

13 Let G be the plane projection of an icosahedron illustrated in Fig. 4.2.
   (a) Exhibit a proper 4-coloring of the vertices of G.
   (b) Prove that Lemma 4.10 cannot be strengthened to the following: If G is a planar graph, then $\delta(G) \leq 4$.

14 Let $G = P_3 \vee C_4$, the join of P_3 and C_4.
   (a) Prove that G is not planar.
   (b) Exhibit G as a bagel graph.

15 Find a graph G that contains a subgraph homeomorphic to K_3, but such that the chromatic number $\chi(G) = 2$.

16 Let G be a plane projection of a regular tetrahedron (a pyramid with a triangular base). Prove the following:
   (a) G is isomorphic to K_4.
   (b) G is isomorphic to G^d.

17 Let $G = K_{3,3}$.
   (a) Prove that $\gamma(G) \leq 1$ by embedding G in a plane with one overpass.
   (b) Explain how part (a) completes the proof that $\gamma(K_{3,3}) = 1$.

18 Let G be a plane graph.
   (a) If G is connected, show that $(G^d)^d = G$.
   (b) If $G = 2K_3$, show that $(G^d)^d$ is isomorphic to the overlap of K_3 and K_3 in K_1.

19 Let G be a plane graph. Prove that G_d is connected.

20 Let Y be the set of edges along some cycle of a connected plane graph G. Denote by Y^d the corresponding set of edges of G^d. Prove that Y^d is a disconnecting set of edges of G^d.

21 The *Heawood graph* H is shown in Fig. 4.13. Prove that $\gamma(H) \leq 1$ by embedding H in an orientable surface of genus 1 (a plane with a single overpass).

22 Let $G = K_{4,4}$.
   (a) Explain why $\gamma(G) \geq 1$.
   (b) Exhibit G as a bagel graph by embedding it in a surface S consisting of a plane with a single overpass.

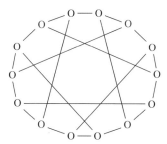

Figure 4.13

(c) Use Euler's Formula (for connected graphs of arbitrary genus) to compute $r(G)$.

(d) Use your illustration from part (b) to confirm your answer to part (c).

23 Show that no complete graph has genus 7.

24 Compute $\chi(\Pi_p)$ for $p =$
 (a) 2. (b) 3. (c) 4. (d) 5. (e) 6.

25 According to Theorem 4.21, K_7 can be embedded in a torus. How many regions would the resulting bagel graph have?

26 Let $G = K_6$.
 (a) Use Theorem 4.21 to show that $\gamma(G) = 1$.
 (b) Exhibit G as a bagel graph by embedding it in the surface obtained by adding a single overpass to the plane.
 (c) Confirm Euler's Formula (for connected graphs of arbitrary genus) by counting regions in your solution to part (b).

27 G. Ringel has shown that $\gamma(K_{s,t}) = \lceil (s-2)(t-2)/4 \rceil$ (unless $s = t = 1$).
 (a) Prove that $K_{2,t}$ is planar using Ringel's formula.
 (b) Prove that $K_{2,t}$ is planar by embedding it in the plane.
 (c) Compute $\gamma(K_{t,t})$ for $t = 4, 5, 6,$ and 7.

28 Let G be a graph on $n \geq 4$ vertices and m edges. Prove that $\gamma(G) \geq \lceil 1 + (m - 3n)/6 \rceil$.

29 Find the error in the following "proof" of the theorem of Ringel and Youngs: Let G be a connected graph on $n \geq 4$ vertices. Assume G has been embedded in a surface of genus $\gamma = \gamma(G)$. Let H be a triangulation of G. If H has m edges, then counting edges around the perimeter of each of the $r = r(H)$ regions yields the identity $2m = 3r$. Let $\bar{d}$ be the average of the vertex degrees of H. By the first theorem of graph theory, $2m = n\bar{d}$. Substitute $r = n\bar{d}/3$ and $m = n\bar{d}/2$ into Equation (27). Rearranging terms

gives $6 + 12(\gamma - 1)/n = \bar{d} \le n - 1$ (with equality if and only if $H = K_n$). Solving the inequality, we obtain $\gamma \le (n-3)(n-4)/12$ which, together with (32), completes the proof of Theorem 4.21.

30. Let A_n be the unique connected, antiregular graph on n vertices. (See Chapter 2, Exercises 36–38.)

    (a) Show that $\gamma(A_7) = 0$.
    (b) Find $\gamma(A_8)$.
    (c) Find $\gamma(A_9)$.
    (d) Show that $\gamma(A_{10}) = 1$.

31. The discovery of higher fullerenes (see Exercise 9) set off a furious race to describe their structures.[13] The search for the most chemically stable configurations precipitated a revival of one of the old war horses of theoretical organic chemistry, the Hückel molecular orbital (HMO) method. By making simplifying approximations in the quantum mechanical Schrödinger equation, Hückel discovered a correlation between chemical stability and graph *energy*. The *energy* of a connected, planar "Hückel" graph G on n vertices is

$$H(G) = \sum_{i=1}^{n} |\gamma_i|,$$

where $\gamma_1, \gamma_2, \ldots, \gamma_n$ are the eigenvalues of the adjacency matrix, $A(G)$. As a rough-and-ready estimate, molecules whose chemical graphs satisfy $H(G) > n$ are predicted to be stable.

   (a) Compute the energy of propylene (chemical formula C_3H_6) whose Hückel graph is the path P_3.
   (b) Compute the energy of benzene (chemical formula C_6H_6) whose Hückel graph is the cycle C_6.

32. Let $G = Q_k$ be the cube graph of Chapter 1, Exercise 33.

    (a) Show that Q_3 is planar.
    (b) Show that Q_4 is not planar.

33. Let G be the plane graph illustrated in Fig. 4.7. Find a nonisomorphic plane graph H on 5 vertices such that $H^d \cong G^d$.

34. Let G be the "cocktail party" graph shown in Fig. 4.14.

    (a) Show that $G \cong (3K_2)^c$.
    (b) Illustrate G as a plane graph.

---

[13] The 24 possibilities for C_{84} are identified in [D.E. Manolopoulos and P. W. Fowler, Molecular graphs, point groups, and fullerenes, *J. Chem. Phys.* **96**(1992), 7603–7614].

Figure 4.14

35 Let G and H be the graphs in Fig. 4.15.
  (a) Show that G is the plane projection of a regular octahedron.
  (b) Show that G is isomorphic to the graph in Fig. 4.14.

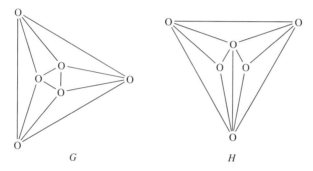

Figure 4.15

  (c) Show that $G^d = G_d$ is isomorphic to the plane projection of a cube.
  (d) Show that G is a triangulation of an appropriate plane drawing of C_6.
  (e) Show that $\chi(G) = 3$.
  (f) Show that H is a triangulation of an appropriate plane drawing of C_6.
  (g) Show that $\chi(H) = 4$.
  (h) Prove that G and H are not isomorphic.[14]

---

[14] It was proved by M. Król [On a sufficient and necessary condition of 3-colorability of a planar graph, I and II, *Prace Nauk. Inst. Mat. Fiz.* **6** (1972), 37–40 and **9** (1973), 49–54] that a planar graph is properly 3-colorable if and only if it is isomorphic to a subgraph of some plane triangulation each of whose vertex degrees is even. Nevertheless, it is an NP-complete problem to decide if a given planar graph is properly 3-colorable. (See [M. R. Garey, D. S. Johnson, and L. J. Stockmeyer, Some simplified NP-complete graph problems, *Theoretical Computer Science* **1** (1976), 237–267].) At the other extreme, H. Grötzsch [Ein Dreifarbensatz für dreikreisfreie Netze auf der Kugel, *Wiss. Z. Martin Luther Univ.* **8** (1959), 109–120] showed that every triangle-free planar graph G (i.e., $\omega(G) \leq 2$) is properly 3-colorable. (See [T. R. Jensen and B. Toft, *Graph Coloring Problems*, Wiley, New York, 1995] for an encyclopedic discussion of such issues).

(i) Show that $p(G, x) > 0$ for all real $x > 4$.[15]

(j) Show that $p(H, x) > 0$ for all real $x > 4$.

---

[15] Attempting to resolve the (then) Four-Color Conjecture, G. D. Birkhoff and D. C. Lewis [Chromatic polynomials, *Trans. Amer. Math. Soc.* **60** (1946), 355–451] proved that, for any plane triangulation G, $p(G, x) > 0$ for all real $x \geq 5$; they conjectured that $p(G, x) > 0$ for all $x \geq 4$. It is still not known whether there exists a plane triangulation G and a real number $x_0 \in (4, 5)$ such that $p(G, x_0) = 0$.

5

Hamiltonian Cycles

Let's plan a trip cycling around Europe. Flying in, say, to Amsterdam we cycle to Berlin, Copenhagen, Danzig, Edinburgh (by ferry), and so on, each leg of the trip taking us to a new city, until returning us to Amsterdam for the flight home. Given two months for the trip, we might hope to visit as many as twenty cities. While it may not be as much as "half the fun," something pleasurable may come just from planning such a trip — or so thought William Rowan Hamilton. In 1857, Hamilton developed a trip-planning *game*, basing it on a regular dodecahedron.

A dodecahedron is a polyhedron having 12 pentagonal faces. Because each pentagon has five vertices, and each vertex is shared by three pentagons, a dodecahedron has $(12 \times 5)/3 = 20$ vertices. Labeling each vertex with the name of a city, Hamilton invited players to find routes, along the edges of the dodecahedron, that visit each city exactly once and end up back at the starting point. Figure 5.1 shows the "Mercator projection" of a regular dodecahedron. Traveling along the edges of this graph, from vertex to vertex in the order indicated, results in what has come to be known as a hamiltonian cycle.[1]

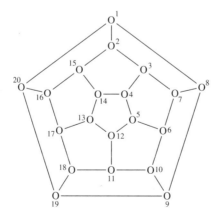

Figure 5.1

---

[1] Two years before Hamilton introduced his game, Thomas Kirkman posed the following problem: Given the graph of a polyhedron, is it always possible to find a cycle containing every vertex?

5.1 Definition. A path containing every vertex of G is a *hamiltonian path*. A cycle containing every vertex is a *hamiltonian cycle*. A graph is *hamiltonian* if it contains a hamiltonian cycle.

A hamiltonian path in a graph on n vertices is a spanning subgraph isomorphic to P_n, and a hamiltonian cycle is a subgraph isomorphic to C_n. Thus, G is hamiltonian if and only if it contains a spanning cycle.[2]

From Fig. 5.1, it is easy to see that the dodecahedron has a spanning cycle. In fact, if we regard two cycles as *equivalent* if they cycle through the same vertices, in the same order (without regard to starting = ending point), then the dodecahedron has 60 inequivalent spanning cycles, occurring in natural (forward–backward) pairs — for example,

⟨1, 2, 3, 4, 5, 6, 7, 8, 9, 10, 11, 12, 13, 14, 15, 16, 17, 18, 19, 20⟩ and
⟨1, 20, 19, 18, 17, 16, 15, 14, 13, 12, 11, 10, 9, 8, 7, 6, 5, 4, 3, 2⟩;
⟨1, 8, 9, 10, 11, 12, 13, 14, 4, 5, 6, 7, 3, 2, 15, 16, 17, 18, 19, 20⟩ and
⟨1, 20, 19, 18, 17, 16, 15, 2, 3, 7, 6, 5, 4, 14, 13, 12, 11, 10, 9, 8⟩;
⟨1, 2, 15, 16, 17, 18, 11, 10, 6, 5, 12, 13, 14, 4, 3, 7, 8, 9, 19, 20⟩ and
⟨1, 20, 19, 9, 8, 7, 3, 4, 14, 13, 12, 5, 6, 10, 11, 18, 17, 16, 15, 2⟩;
⟨1, 20, 16, 15, 2, 3, 4, 14, 13, 17, 18, 19, 9, 10, 11, 12, 5, 6, 7, 8⟩ and
⟨1, 8, 7, 6, 5, 12, 11, 10, 9, 19, 18, 17, 13, 14, 4, 3, 2, 15, 16, 20⟩;

and so on.

Figure 5.2 contains two illustrations of the Petersen graph. (The proof that these two different looking graph are isomorphic is "by the numbers.") Unlike Fig. 5.1, the numbers in Fig. 5.2 do *not* exhibit a spanning cycle. Indeed, the point of the figure is to facilitate a discussion of whether the Petersen graph has a hamiltonian cycle. It is clear from the right-hand picture that it has a cycle of length 9, but what about a cycle of length 10? Since we may "start" a spanning cycle at any vertex, it may as well be with vertex 1. There will be three

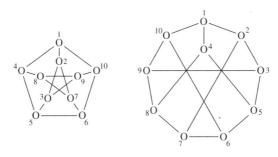

Figure 5.2. Two illustrations of the Petersen graph.

---

[2] Not to be confused with an induced cycle.

choices for the second vertex, either 10, 2, or 4. Because vertices 10 and 2 are symmetrically placed (in the right-hand picture), we may eliminate one of them, say 10, from considerations of existence. Searching for a 10-cycle, suppose we start with the path [1, 4, 5, 6, 7, 2, 3, 9]. If we go to 8 next, we're stuck with no place to exit because the neighbors of vertex 8 have already been visited. If the next vertex were 10, then, because 6 and 9 have already been visited, we would have no choice but to cycle back to 1, leaving out vertex 8; that is, $\langle 1, 4, 5, 6, 7, 2, 3, 9, 10 \rangle$ is another 9-cycle in G. Maybe starting with [1, 4, 5, 3] would be better. From 3 we might, for example, patch on [3, 2, 7] unless, perhaps, [3, 9, 8, 7] seems more promising.

Eventually, by exhausting all possibilities, one finds that the Petersen graph is not hamiltonian.[3] It does not contain a subgraph isomorphic to C_{10}. Of course, few things stimulate mathematical thinking more than the prospect of having to exhaust a large number of cases.

Let G be a graph on n vertices. Suppose H is a fixed but arbitrary spanning subgraph of G. Let S be a vertex cut of G. Because it is a spanning subgraph of $G - S$, the number of connected components of $H - S$ can be no smaller than the number of components of $G - S$. Therefore, $\kappa(G) \geq \kappa(H)$. It follows that $\kappa(G) \geq \kappa(C_n) = 2$ for any hamiltonian graph G. This proves the following.

5.2 Theorem. *Every hamiltonian graph is 2-connected.*

In fact, a little more can be squeezed from this idea. If $u \in V(C_n)$, then $C_n - u \cong P_{n-1}$ is connected. Suppose $v \in V(C_n - u)$. If v is a pendant vertex of $C_n - u$, then $C_n - u - v \cong P_{n-2}$ is still connected; otherwise, it is a graph with two components, each of which is isomorphic to a path. Removing a third vertex results in a graph with at most three components, and so on. In general, the number of components of $C_n - S$ is at most o(S) whenever $\phi \neq S \subset V(C_n)$. These observations prove the following.

5.3 Theorem. *If G is hamiltonian, then the number of components of $G - S$ is at most the cardinality of S, $\phi \neq S \subset V(G)$.*

Let $\xi(G)$ be the number of components[4] of G. By Theorem 5.3, G is hamiltonian only if $\xi(G - S) \leq$ o(S), $\phi \neq S \subset V(G)$.

5.4 Example. Let G be the 2-connected "frog graph" in Fig. 5.3. If $S = \{u, v\}$, then $\xi(G - S) = 3 > 2 =$ o(S). Hence, by Theorem 5.3, G is not hamiltonian. $\square$

---

[3] Computers are useful for this sort of thing. For the Petersen graph, $2 \times 3^8 = 13{,}122$ bounds the number of possibilities, so almost any algorithm should produce a negative answer in a reasonable amount of time. On the other hand, if we really needed to know about some graph with hundreds of vertices, it might be worth searching for a good algorithm to find hamiltonian cycles. How hard should one look? This is what computational complexity is all about. One of the reasons the hamiltonian cycle problem is of interest is that it is known to be NP-complete.

[4] Don't confuse $\xi(G)$ with the edge connectivity, $\varepsilon(G)$.

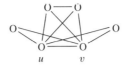

Figure 5.3

What about the converse of Theorem 5.3? Might this necessary condition also be sufficient? No — the Petersen graph is a counterexample. We must look elsewhere for sufficient conditions.

5.5 Ore's Lemma.[5] *Let G be a graph on n vertices. Suppose u and w are two nonadjacent vertices of G such that $d_G(u) + d_G(w) \geq n$. Let $G_1 = G + uw$, the graph obtained from G by adding a new edge $e = uw$. Then G is hamiltonian if and only if G_1 is hamiltonian.*

Proof. Necessity is obvious. Adding an edge cannot destroy an existing cycle.

Conversely, if G_1 is hamiltonian but G is not, then G must contain a spanning path $[u = v_1, v_2, \ldots, v_n = w]$. Let $A = \{i : v_i w \in E(G)\}$ and $B = \{i : uv_{i+1} \in E(G)\}$. Then $n \notin A$ because $v_n = w$, and $n \notin B$ because there is no v_{n+1}. Since $o(A) = d_G(w)$, $o(B) = d_G(u)$, and $n \notin A \cup B$, we have

$$n - 1 \geq o(A \cup B)$$
$$= o(A) + o(B) - o(A \cap B)$$
$$= d_G(u) + d_G(B) - o(A \cap B)$$
$$\geq n - o(A \cap B),$$

proving that $A \cap B \neq \phi$. So, there is an integer $k \leq n - 1$ such that $v_k w \in E(G)$ and $uv_{k+1} \in E(G)$. Since $uw \notin E(G)$, $1 < k < n - 1$. Therefore, $\langle u = v_1, v_2, \ldots, v_k, w = v_n, v_{n-1}, \ldots, v_{k+1}\rangle$ is a spanning cycle of G. ∎

The following definition is motivated by Ore's Lemma.

5.6 Definition. Let G be a graph with vertex set $V(G) = \{v_1, v_2, \ldots, v_n\}$. The *closure* of G, denoted $\overline{G}$, is the graph generated by the following algorithm:

1. Let $H = G$.
2. For $i = 1$ to $n - 1$
3. For $j = i + 1$ to n
4. If $v_i v_j \notin E(G)$ and $d_G(v_i) + d_G(v_j) \geq n$, then $G = G + v_i v_j$.
5. Next j

[5] O. Ore, Note on Hamilton circuits, *Amer. Math. Monthly* **67** (1960), 55.

6. Next i
7. If $G \neq H$, then Step 1.
8. $\overline{G} = G$.

In ordinary English prose, $\overline{G}$ is obtained recursively by adding new edges between nonadjacent vertices, the sum of whose degrees is not less than n, until no such pair remains.

5.7 Example. As illustrated in Fig. 5.4, the closure of $G = C_5 + e$ is $\overline{G} = K_5$. □

Figure 5.4

5.8 Theorem. *Let G be a graph. Then G is hamiltonian if and only if $\overline{G}$ is hamiltonian.*

Proof. Immediate from Ore's Lemma and the definitions. ∎

5.9 Corollary. *Let G be a graph on $n \geq 3$ vertices. If the closure of G is complete, then G is hamiltonian.*

Proof. Immediate from Theorem 5.8 and the fact that K_n is hamiltonian for all $n \geq 3$. ∎

5.10 Corollary. *Let G be a graph on $n \geq 3$ vertices. If its minimum vertex degree, $\delta(G) \geq n/2$, then G is hamiltonian.*

Proof. Immediate from Corollary 5.9 and the definitions. ∎

There is a technical difficulty with these results. What is there to guarantee that $\overline{G}$ is well-defined? I might recursively add edges between nonadjacent pairs of vertices ... and arrive at $\overline{G} = H_1$. Meanwhile, joining different nonadjacent pairs, or even some of the same pairs but in a different order, you might end up with $\overline{G} = H_2$. If $H_1 \neq H_2$, the last few paragraphs will need to be revised.

5.11 Theorem. *The closure of G is well-defined.*

Proof. Suppose $H_1 = G + e_1 + e_2 + \cdots + e_r$ and $H_2 = G + e'_1 + e'_2 + \cdots + e'_s$ are closures of G. We claim that $e_i \in E(H_2)$, $1 \leq i \leq r$, and $e'_j \in E(H_1)$,

$1 \leq j \leq s$. Otherwise, without loss of generality we may assume k is the smallest positive integer such that $e_k \notin E(H_2)$. Let $e_k = uv$ and define $H = G + e_1 + \cdots + e_{k-1}$. It follows from the derivation of H_1 that $d_H(u) + d_H(v) \geq n$. By the choice of e_k, H is a subgraph of H_2. Therefore, the sum of the degrees of u and v in H_2 can be no less than n, meaning that e_k must be an edge of the closure H_2. This contradiction establishes the claim and proves the theorem. ∎

5.12 Example. Corollary 5.9 gives a sufficient condition for a graph to be hamiltonian. Is the condition also necessary? Is G hamiltonian only if $\overline{G} = K_n$? Answer: No. If $G = C_5$ then G is hamiltonian, but $\overline{G} = G \neq K_5$. □

Recall that $d(G) = (d_1, d_2, \ldots, d_n)$ where $d_1 \geq d_2 \geq \cdots \geq d_n$ are the degrees of the vertices of G arranged in nonincreasing order. The next result involves a condition on $d(G)$ sufficient to guarantee that G is hamiltonian. It turns out that the condition is more easily stated when the degrees are rearranged in nondecreasing order. For this purpose, define $\delta_i = d_{n-i+1}$, $1 \leq i \leq n$. Then $\delta_1 = d_n$, $\delta_2 = d_{n-1}, \ldots, \delta_n = d_1$, and $\delta_1 \leq \delta_2 \leq \cdots \leq \delta_n$.

5.13 Pósa's Theorem.[6] *Let G be a graph on $n \geq 3$ vertices of degrees $\delta_1 \leq \delta_2 \leq \cdots \leq \delta_n$. If $\delta_k > k$, $1 \leq k < n/2$, then G is hamiltonian.*

Because its hypotheses are satisfied when $\delta(G) \geq n/2$, Pósa's Theorem is stronger (weaker hypotheses, same conclusion) than Corollary 5.10. We are, in fact, going to prove an even stronger result.

5.14 (Technical) Lemma.[7] *Let G be a graph on $n \geq 3$ vertices of degrees $\delta_1 \leq \delta_2 \leq \cdots \leq \delta_n$. If $\delta_k > k$ or $\delta_{n-k} \geq n - k$, $1 \leq k < n/2$, then G is hamiltonian.*

The value of Lemma 5.14 lies not in its simplicity, but in its usefulness. Apart from being stronger than Pósa's Theorem, it is the key to Chvátal's Theorem (below). On the other hand, even this complicated sufficient condition is not necessary. If $G = C_7$, then $2 = \delta_3 < 3$ and $2 = \delta_4 < 4$.

Proof. We will use the hypotheses of Lemma 5.14 to prove that $\overline{G} = K_n$. The proof is by contradiction. If $\overline{G} \neq K_n$, then $\overline{G}$ has a pair of nonadjacent vertices. From all such nonadjacent pairs, choose u and w so that $\overline{d}(u) + \overline{d}(w)$ is as large as possible, where $\overline{d}(v) = d_{\overline{G}}(v)$ denotes the degree of vertex v in $\overline{G}$. Let $p = \overline{d}(u)$ and $q = \overline{d}(w)$. Without loss of generality, we may assume $p \leq q$. Moreover, $p + q \leq n - 1$; otherwise $uw \in E(\overline{G})$. In particular, $p < n/2$.

Apart from w itself, there exist $(n - 1) - q$ vertices of $\overline{G}$ that are not adjacent to w. If v is any one of them, then, by our choice of u and w, $\overline{d}(v) \leq \overline{d}(u) = p$. Hence, at least $n - 1 - q$ of the vertices of $\overline{G}$ have degrees at most p; that

---

[6] L. Pósa, A theorem concerning Hamilton lines, *Magyar Tud. Akad. Mat. Kutató Int. Kozl.* **7** (1962), 225–226.
[7] Lemma 5.14 appeared in [V. Chvátal, On Hamilton's ideals, *J. Combinatorial Theory B* **12** (1972), 163–168].

is, $\bar{\delta}_1 \le \bar{\delta}_2 \le \cdots \le \bar{\delta}_{n-1-q} \le p$. Because $p \le n - 1 - q$, we see that $\delta_p \le \bar{\delta}_p \le \bar{\delta}_{n-1-q} \le p$. Since $p < n/2$, the hypotheses guarantee that $n - p \le \delta_{n-p} \le \bar{\delta}_{n-p}$. Thus, to reach a contradiction, it suffices to show $\bar{\delta}_{n-p} \le q$. (Why?)

Apart from u itself, there exist $(n - 1) - p$ vertices of $\overline{G}$ that are not adjacent to u. If v is any one of them, then $\bar{d}(v) \le \bar{d}(w) = q$. Because $\bar{d}(u) = p \le q$, at least $n - p$ vertices of $\overline{G}$ have degrees at most q; that is, $\bar{\delta}_1 \le \bar{\delta}_2 \le \cdots \le \bar{\delta}_{n-p} \le q$. ∎

Denote the graph in Fig. 5.5 by (the descriptive name) λ.

Figure 5.5

5.15 Theorem. *Let G be a 2-connected graph. If no induced subgraph of G is isomorphic to $K_{1,3}$ or to λ, then G is hamiltonian.*

Proof. Because G is 2-connected, it must have at least three vertices and at least one cycle. (Indeed, by Theorem 3.15, every pair of vertices of G lies on a cycle.) Let C be a longest cycle of G. If C is not a spanning cycle, then there is a vertex v on C and a vertex u not on C such that $uv \in E(G)$. Let x and y be the vertices of C adjacent to v. If $ux \in E(G)$, then replacing the subpath $[x, v]$ of C with the path $[x, u, v]$ produces a longer cycle. Similarly, $uy \notin E(G)$. Let $H = G[u, v, x, y]$, the subgraph of G induced by $\{u, v, x, y\}$. If $xy \in E(G)$, then $H \cong \lambda$. Otherwise, $H \cong S_4 = K_{1,3}$. ∎

V. Chvátal found an interesting generalization of the graph λ appearing in Theorem 5.15.

5.16 Definition. If $n \ge 3$ and $1 \le r < n/2$, the *Chvátal graph* $G_{n,r} = K_r \vee (K_r^c \oplus K_{n-2r})$.

5.17 Example. The Chvátal graph $G_{4,1} = K_1 \vee (K_1 \oplus K_2) \cong \lambda$, the forbidden subgraph from Theorem 5.15; $G_{7,2} = K_2 \vee (K_2^c \oplus K_3)$ is illustrated in Fig. 5.6. In general, the degree sequence of $G_{n,r}$ is

$$d(G_{n,r}) = (n - 1, \ldots, n - 1, n - r - 1, \ldots, n - r - 1, r, \ldots, r), \quad (34)$$

where $n - r - 1$ occurs with multiplicity $n - 2r$, and both $n - 1$ and r occur with multiplicity r. □

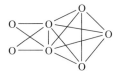

Figure 5.6

Let S be the set of vertices of $G_{n,r}$ corresponding to K_r. Then $\xi(G_{n,r} - S) = r + 1 > r = o(S)$ so, by Theorem 5.3, $G_{n,r}$ is not hamiltonian. In the sense about to be defined, the Chvátal graphs $G_{n,r}$, $1 \leq r < n/2$, are the dominant nonhamiltonian graphs on n vertices.

5.18 Definition. Let $a_1 \geq a_2 \geq \cdots \geq a_n$ and $b_1 \geq b_2 \geq \cdots \geq b_n$ be two nonincreasing integer sequences. If $a_i \geq b_i$, $1 \leq i \leq n$, then the sequence of a's is said to *dominate* the sequence of b's.

5.19 Chvátal's Theorem. *Let G be a nonhamiltonian graph on $n \geq 3$ vertices. Then there exists a positive integer $r < n/2$ such that $d(G_{n,r})$ dominates $d(G)$.*

Proof. Let $\delta_1 \leq \delta_2 \leq \cdots \leq \delta_n$ be the degree sequence of G in nondecreasing order. If $\delta_k > k$ or $\delta_{n-k} \geq n - k$, for all $k < n/2$, then, by Lemma 5.14, G is hamiltonian. Thus, there must exist a positive integer $r < n/2$ such that $\delta_r \leq r$ and $\delta_{n-r} \leq n - r - 1$; that is, $\delta_i \leq r$, $1 \leq i \leq r$ and $\delta_i \leq n - r - 1$, $r + 1 \leq i \leq n - r$. Because $\delta_i \leq n - 1$ for all i, it follows from Equation (34) that $d(G_{n,r})$ dominates $d(G)$. ∎

Chvátal's Theorem gives another sufficient condition for a graph G on n vertices to be hamiltonian, namely, if $d(G)$ is not dominated by $d(G_{n,r})$ for any $r < n/2$. The next sufficient condition is as elegant as it is unexpected.

5.20 Theorem. *Let G be a graph on $n \geq 3$ vertices. If the connectivity of G is an upper bound for its independence number — that is, if $\kappa(G) \geq \alpha(G)$ — then G is hamiltonian.*

Proof. If $\alpha(G) = 1$, then $G = K_n$, a hamiltonian graph. So, we may assume $k = \kappa(G) \geq \alpha(G) \geq 2$. By Theorem 3.20, every set of k vertices of G lies on a cycle. Let C be a cycle of maximum length $\ell \geq k$. If $\ell = n$, the proof is complete. Otherwise, let w_i, $1 \leq i \leq k$, be distinct vertices of C and u a vertex of G not on C. By Corollary 3.19, there exist paths $P_1, P_2, \ldots, P_k$ in G such that

$$P_i \text{ is a path from } u \text{ to } w_i, \quad 1 \leq i \leq k,$$

and (35)

$$P_i \cap P_j = \{u\} \quad \text{whenever } i \neq j.$$

Chap. 5 Hamiltonian Cycles

If $P_i = [u, v_1, \ldots, v_r, w_i]$, it may be that one of the v's is a vertex of C. If so, let t be the smallest integer such that v_t lies on C. If we replace P_i with $[u, v_1, \ldots, v_t]$ and rename vertices so that v_t becomes (the new) w_i, conditions (35) remain valid. Thus, we may assume that $P_i \cap C = \{w_i\}$, $1 \le i \le k$, and, without loss of generality, that $w_1, w_2, \ldots, w_k$ are arranged consecutively around C. Finally, it is convenient to define $w_{k+1} = w_1$.

If w_1 and w_2 were adjacent vertices along C, a cycle longer than C could be obtained by replacing subpath $[w_1, w_2]$ with $P_1^{-1} P_2$. Evidently, because C is a longest cycle, w_i and w_{i+1} cannot be consecutive vertices of C for any i. Let x_i be the vertex of C adjacent to w_i on the way to w_{i+1} as in Fig. 5.7. If $ux_i \in E(G)$ for some i, a cycle longer than C could be obtained by replacing the subpath $[w_i, x_i]$ with $P_i^{-1} x_i$.

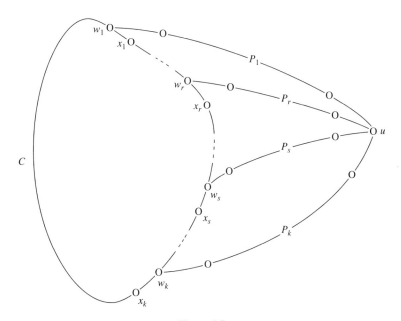

Figure 5.7

Consider the set $S = \{u, x_1, x_2, \ldots, x_k\}$. Because $o(S) > k \ge \alpha(G)$ and $ux_i \notin E(G)$ for all i, it must be that $\{x_r, x_s\} \in E(G)$ for some r and s satisfying $1 \le r < s \le k$. We can now produce a cycle C' from C by replacing the subpath

$$[w_r, x_r, \ldots, w_s, x_s]$$

with a path from w_r to x_s constructed as follows: Go from w_r to u backwards along path P_r; from u to w_s along P_s; from w_s to x_r backwards along C; and from x_r to x_s along $[x_r, x_s]$. Because it contains u in addition to all the vertices of C, C'

is longer than a longest cycle of G. This contradiction proves the nonexistence of u; that is, a longest cycle of G is a spanning cycle. ∎

Back in the 1930s, Hassler Whitney proved the following theorem about what was then the Four-Color Conjecture.

5.21 Whitney's Theorem.[8] *Every planar graph can be colored properly with four colors, if and only if every planar hamiltonian graph can be colored properly with four colors.*

Let's have a look at hamiltonian plane graphs. Let G be the plane projection of a regular dodecahedron exhibited on the right-hand side of Fig. 5.8. The hamiltonian cycle C, on the left-hand side, came from our discussion of Fig. 5.1. Note that the complement of C in the plane consists of two connected components, one *inside* the curve and the other *outside*.[9] Let e be an edge of G that does not lie along C. Since G is a plane graph, C meets the arc corresponding to e only at the points representing the vertices of G incident with e. Call the arc representing e a *chord* of C. Then, apart from its endpoints, every chord lies entirely inside or entirely outside C.

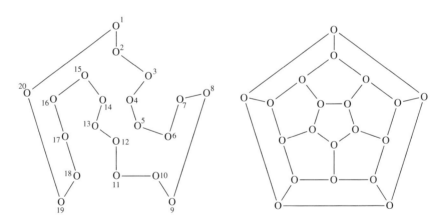

Figure 5.8

What happens if we add a chord of C to the left-hand side of Fig. 5.8? Adding an inside chord, say $\{v_4, v_{14}\}$, partitions the interior of C into two regions. Adding chord $\{v_1, v_8\}$ divides the exterior into two regions. Every time a chord is added,

---

[8] See H. Whitney, A theorem on graphs, *Annals of Math.* **32** (1931), 378–390. Due to this result, Francis Guthrie's innocent problem became a motivating force in the development of hamiltonian graph theory.

[9] Strictly speaking, this is a consequence of the *Jordan Curve Theorem*. See, e.g., J. R. Munkres, *Topology, a first course*, Prentice-Hall, Englewood Cliffs, NJ, 1975.

Chap. 5 Hamiltonian Cycles

the number of regions goes up by one. Because chords cannot cross each other, adding an inside chord must split an existing interior region in two; adding an outside chord splits an exterior region in two.

More generally, suppose G is a hamiltonian plane graph on n vertices. Let C be a hamiltonian cycle. If C has s inside chords and t outside chords, then G has a total of $m = n + s + t$ edges, $s + 1$ of the regions of G are inside C, and $t + 1$ of them are outside.[10] Carried a little further, this analysis yields the following.

5.22 Grinberg's Theorem.[11]   *Suppose C is a hamiltonian cycle in the plane graph G. Let s_i and t_i be the numbers of regions of G inside and outside of C (respectively), the perimeters of which contain exactly i edges. Then*

$$\sum_{i=3}^{n}(i-2)(s_i - t_i) = 0. \tag{36}$$

Proof.   Let s and t be the numbers of chords that are inside and outside of C, respectively. Then, by our previous discussion, $s + 1 = \sum s_i$ and $t + 1 = \sum t_i$.

If we assign to each interior region the number of edges in its perimeter, then the sum of these numbers is $\sum i s_i$. In this total, each of the s inside chords is counted twice while each edge of C is counted once. Thus, $\sum i s_i = 2s + n$. Substituting $s = -1 + \sum s_i$ in this expression, and rearranging terms, we obtain

$$\sum_{i=3}^{n}(i-2)s_i = n - 2. \tag{37}$$

A similar calculation involving the exterior regions produces

$$\sum_{i=3}^{n}(i-2)t_i = n - 2. \tag{38}$$

Subtracting (38) from (37) yields the result. ∎

5.23 Example.   The boundary cycle of every region in the dodecahedron graph illustrated on the right-hand side of Fig. 5.8 consists of 5 edges. Thus, when the right- and left-hand sides of Fig. 5.8 are superimposed, $s_i = t_i = 0$, $i \neq 5$. With a little care (check it!) one finds that $s_5 = t_5 = 6$. In this case, Equation (37) becomes

$$1 \times 0 + 2 \times 0 + 3 \times 6 + 4 \times 0 + \cdots + 18 \times 0 = 20 - 2,$$

while Equation (36) reduces to

$$1 \times (0 - 0) + 2 \times (0 - 0) + 3 \times (6 - 6) + 4 \times (0 - 0) + \cdots = 0,$$

---

[10] Confirm that these values are consistent with Euler's Formula for plane graphs.
[11] E. J. Grinberg, Plane homogeneous graphs of degree three without hamiltonian circuits, *Latvian Math. Yearbook* **4** (1968), 51–58. (Also see G. Chartrand and L. Lesniak, *Graphs & Digraphs*, 2nd ed., Wadsworth & Brooks/Cole, Belmont, CA, 1986, p. 197.)

providing a nice illustration of Grinberg's Theorem. On the other hand, it follows from Grinberg's Theorem that, no matter what hamiltonian cycle of the dodecahedron graph one superimposes on the right-hand side of Fig. 5.8, exactly half of the 12 regions (corresponding to the 12 faces of the dodecahedron) must be inside the cycle, and the other half outside! □

In view of Whitney's Theorem, the following result must have been very tantalizing to those attempting to prove the Four-Color Conjecture.

5.24 Theorem. *If G is a hamiltonian plane graph, then its dual multigraph is properly 4-colorable; that is, $\chi(G^d) \leq 4$.*

Proof. Let C be a hamiltonian cycle in the plane graph G. An edge of C lies on the boundary of two regions of G if and only if one of the regions is inside C and the other is outside. If the perimeters of two inside regions of C share an edge, the arc representing that edge is an interior chord of C. Therefore, no two inside regions of C have perimeters that share as many as two edges.

Place one dot in the interior of each inside region of G. Denote the set of these dots by D_I. If the boundary cycles of two inside regions share an edge of G, draw an arc that connects the corresponding dots and crosses (only) the edge of G shared by the two boundaries. Denote the set of these arcs by A_I. Because G is 2-connected, it has no bridges. Thus, no arc of A_I is a loop from a dot to itself. Because the boundary cycles of inside regions share at most one edge, no two arcs of A_I connect the same two dots. In other words, $H_I = (D_I, A_I)$ is a (plane) graph, lying entirely inside C. If H_I had a cycle, that cycle would have an interior, and that interior would contain at least one vertex of G, contradicting the fact that every vertex of G lies on C. Having no cycles, H_I has no odd cycles; that is, H_I is bipartite. Its vertices can be colored properly with two colors, say red and white. A similar argument applied to H_E, the geometric dual of the regions of G outside C, shows that its vertices can be colored properly with two colors, say blue and green. Because arcs representing the remaining edges of G^d join an outside dot to an inside dot, the natural 4-coloring of $H_I \oplus H_E$ obtained from the proper colorings of H_I and H_E extends to a proper 4-coloring of G^d. ■

5.25 Corollary. *If G is a hamiltonian plane graph, then $\chi(G_d) \leq 4$.*

Proof. Because $\chi(G_d) = \chi(G^d)$, the result is an immediate consequence of Theorem 5.24. ■

5.26 Example. Let G be the "bow tie" graph illustrated in Fig. 5.9. Then $G_d \cong P_3$. By inspection (it has a cut-vertex), G is not hamiltonian. Might there be a *hamiltonian* plane graph H with $H_d \cong P_3$?

Let C be a spanning cycle in a 3-region, hamiltonian plane graph H. Then at least one region of H is inside C and at least one (the unbounded region) is

Chap. 5 Hamiltonian Cycles

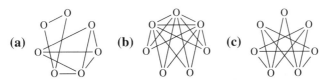

Figure 5.9

outside. If the third region is inside C, then the two interior regions share exactly one boundary edge (corresponding to a chord of C). Moreover, the boundary of each interior region contains an edge of C, which it shares with the perimeter of the outside region. Evidently, in this case, $H_d \cong K_3$. A similar argument shows that $H_d \cong K_3$ when two of the three regions of H are outside C. In particular, $H_d \not\cong P_3$ for any hamiltonian plane graph H. It is because of such "exceptions" that Corollary 5.25 does not prove the Four-Color Theorem.

It is easy to see, of course, that the exceptional graph P_3 is properly 4-colorable. (It is bipartite!) If a simple proof exists that all exceptional plane graphs are properly 4-colorable, it has not been found. □

EXERCISES

1. Determine whether G is hamiltonian and justify your answer.

   (a)   (b)   (c)

2. Show that the conclusion of Ore's Lemma does not follow from the weaker hypothesis $d(u) + d(v) \geq n - 1$.

3. Explain why Theorem 5.2 is a corollary of Theorem 3.15.

4. Show that the complete bipartite graph
   (a) $K_{3,3}$ has six different (forward-backward) pairs of hamiltonian cycles.
   (b) $K_{3,4}$ doesn't have any hamiltonian cycles.
   (c) $K_{r,s}$ is hamiltonian if and only if $r = s > 1$.

5. Let H be a graph on $n \geq 2$ vertices. Suppose $(n-1)/2 \leq \delta(H)$. Prove that H contains a hamiltonian path.

6. Let H be a graph on n vertices. Suppose $\delta_i \geq \delta_i^c$, $1 \leq i \leq n/2$, where $\delta_1^c \leq \delta_2^c \leq \cdots \leq \delta_n^c$ are the vertex degrees of H^c. Prove that H has a hamiltonian path.

7. Prove that every self-complementary graph has a hamiltonian path.

8. Rewrite Pósa's Theorem in terms of d_k (as opposed to δ_k).

9 Let G be a graph with m edges and $n \geq 3$ vertices. Suppose $2m \geq n^2 - 3n + 6$. Prove that G is hamiltonian.

10 Show that it is possible to color one-third of the edges of K_7 red, one-third of them white, and one-third of them blue, in such a way that each of the three uniformly colored sets of edges corresponds to a hamiltonian cycle.

11 Compute the number m of edges of the Chvátal graph
   (a) $G_{9,1}$. (b) $G_{9,2}$. (c) $G_{9,3}$. (d) $G_{9,4}$.

12 Consider the Chvátal graph $G = G_{n,r}$, where $r < n/2$.
   (a) Find the number $m(n, r)$ of edges of G as a function of n and r.
   (b) Compute $M(n) = \max m(n, r)$, as r ranges over the integers in the interval $[1, n/2)$.
   (c) Prove that G is hamiltonian for any graph G on n vertices and $m > M(n)$ edges.

13 Exhibit a hamiltonian graph that contains an induced subgraph
   (a) isomorphic to $S_4 = K_{1,3}$.
   (b) isomorphic to the graph λ in Theorem 5.15.

14 Let A_n be the unique connected antiregular graph on n vertices. (See Chapter 2, Exercise 36.) Prove the following:
   (a) A_4 is the Chvátal graph $G_{4,1}(\cong \lambda)$.
   (b) A_n is not hamiltonian.
   (c) A_n has a hamiltonian path.

15 Recall that two cycles of G are equivalent if they cycle through the same vertices in the same order (without regard to starting = ending point). How many inequivalent hamiltonian cycles does K_5 have?

16 Bill is a traveling salesman whose territory consists of five cities, one of which is his home base. On each business trip, Bill visits every city in his territory (once) before returning home. Figure 5.10 shows the costs (in units of $100) of flying between the cities in Bill's territory.

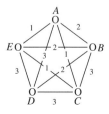

Figure 5.10

(a) Find the minimum airline cost associated with one of Bill's business trips.

(b) Find all the hamiltonian cycles that achieve this minimum cost.

17 Find a nonhamiltonian graph G with the property that $G - v$ is hamiltonian for all $v \in V(G)$.

18 A graph G is *hamiltonian connected* if, for every pair of distinct vertices $u, v \in V(G)$, there is a hamiltonian path in G from u to v.

(a) Show that C_4 is hamiltonian but not hamiltonian connected.

(b) Suppose G is a hamiltonian connected graph on $n \geq 3$ vertices. Prove that G is hamiltonian.

(c) Let G be a graph on $n \geq 3$ vertices. Denote by $\hat{G}$ the graph obtained from G by recursively adding new edges between nonadjacent pairs of vertices the sum of whose degrees is at least $n + 1$. If $\hat{G} = K_n$, prove that G is hamiltonian connected.

19 Let G be a hamiltonian connected graph on $n \geq 4$ vertices. (See Exercise 18 for the definition of hamiltonian connected.) Prove that G is 3-connected.[12]

20 Let $G = (V, E)$ be a bipartite graph, on $n \geq 4$ vertices, and with bipartition $V = V_1 \cup V_2$, where $o(V_1) = o(V_2)$. If $\delta(G) > n/4$, prove that G is hamiltonian.

21 Let G be the plane graph illustrated in Fig. 5.11. Confirm Grinberg's Theorem for G with respect to the hamiltonian cycle $C = \langle v_1, v_2, v_3, v_4, v_5, v_6, v_7, v_8 \rangle$.

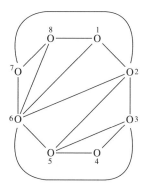

Figure 5.11

---

[12] It was proved by W. T. Tutte [A theorem on planar graphs, *Trans. Amer. Math. Soc.* **82** (1956), 99–116] that every 4-connected planar graph is hamiltonian connected.

22 Let G be a graph on $n \geq 3$ vertices. If G has a hamiltonian path, show that $\xi(G - S) \leq o(S) + 1$ for every proper subset S of $V(G)$.

23 Of the 148 nonisomorphic graphs with 7 vertices and 10 edges, 132 are connected. Of these 132 connected graphs, 30 are hamiltonian. Of these 30 hamiltonian graphs, 29 are planar. Exhibit the unique nonplanar hamiltonian graph having 7 vertices and 10 edges.

24 List all hamiltonian cycles of the graph in Fig. 5.12 that begin with subpath $[v_1, v_2]$.

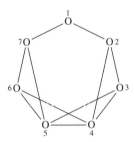

Figure 5.12

25 Show that the graph λ in Fig. 5.5 is isomorphic to

(a) an overlap of K_3 and K_2 in K_1.

(b) $S_4 + e$, the graph obtained from $K_{1,3}$ by adding a new edge joining two of its pendant vertices.

26 Prove that the cube graph Q_k (of Chapter 1, Exercise 33) is hamiltonian, $k \geq 2$.

27 Prove the following:

(a) $G = K_2 \vee (K_1 \oplus K_2 \oplus K_3)$ is not hamiltonian.

(b) $H = K_3 \vee 4K_5$ is not hamiltonian.

28 Theorem 5.19 expresses Chvátal's Theorem in the symbolic form $H \Rightarrow C$, where the hypothesis $H = [G$ is not hamiltonian]. The *negation* of H is $\sim H = [G$ is hamiltonian]. The *contrapositive* form of Chvátal's Theorem, expressed symbolically as $\sim C \Rightarrow \sim H$, is the logically equivalent statement that $\sim C$ is a sufficient condition for G to be hamiltonian.

(a) Express $\sim C$, the negation of C, in words.

(b) Show that $\sim C$ is not a necessary condition for G to be hamiltonian; that is, find a counterexample to the *converse* of Chvátal's Theorem, a symbolic version of which is $\sim H \Rightarrow \sim C$.

Chap. 5 Hamiltonian Cycles

29 Let H be the graph illustrated in Fig. 5.13(a).

(a) Show that H is a plane triangulation of one of its hamiltonian cycles.

(b) Show that $V(H)$ can be partitioned into the disjoint unions of two sets, X and Y, such that the induced subgraphs $H[X]$ and $H[Y]$ are forests.

(c) Show that $H^d = H_d$ is isomorphic to the graph illustrated in Fig. 5.13(b).

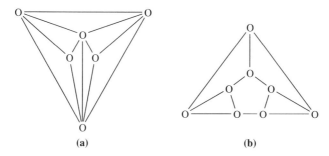

Figure 5.13

(d) Show that H_d is hamiltonian.[13]

30 Let G be a graph having n vertices and m edges. Recall that a *walk* in G of length k, from v to w, is a sequence $v = u_0, u_1, \ldots, u_k = w$ of vertices of G such that $u_{i-1}u_i \in E(G)$, $1 \le i \le k$. A walk of length m, in which every edge of G occurs exactly once, is called an *Euler tour*. (An Euler tour is the edge analog of a hamiltonian path.) Let $H = K_4 - e$ be the graph illustrated in Fig. 5.14.

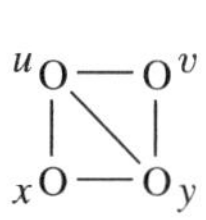

$$\begin{array}{c} {}^u\!\circ\!\!-\!\!\circ^{\,v} \\ |\,\diagdown\!| \\ {}_x\!\circ\!\!-\!\!\circ_{\,y} \end{array}$$

Figure 5.14

(a) Find an Euler tour in H from vertex u to vertex y.

(b) Show that there is no Euler tour in H from vertex v to vertex x.

31 Like many European towns, historic Königsberg (present-day Kaliningrad) began life on an island in a river. By the mid-1700s, when native son

---

[13] It was proved by S. K. Stein [B-sets and planar maps, *Pacific J. Math.* **37** (1971), 217–224] that the vertex set of a plane triangulation G can be partitioned as $V(G) = X \cup Y$, where $G[X]$ and $G[Y]$ are forests, if and only if G_d is hamiltonian.

Immanuel Kant began teaching philosophy there, the town had spilled over to a second island and to both banks of the Pregel. (See Fig. 5.15.) It seems (or, so the story goes) that the prosperous burghers wanted to show the town off by means of a scenic walking tour. They wanted a tour that would lead visitors across each of the seven bridges — exactly once.

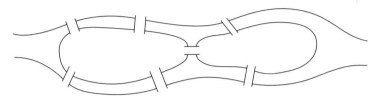

Figure 5.15. Diagram of eighteenth-century Königsberg.

(a) Show that such a scenic tour of Königsberg is equivalent to an Euler tour (Exercise 30) in the multigraph of Fig. 5.16.

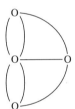

Figure 5.16

(b) Show that an Euler tour in a multigraph M is equivalent to an Euler tour in the graph G obtained by subdividing each edge of M.

(c) Suppose v is a vertex of G of odd degree. Show that any Euler tour in G that does not begin at v must end at v.

(d) Show that a graph with more than two vertices of odd degree cannot have an Euler tour.[14]

32 The island Republic of Vanuatu, formerly known as New Hebrides, lies about 1200 miles east of Australia. The native Malekula peoples illustrate myths about passages to *the Land of the Dead* by means of so-called unicursal diagrams (closed curves that can be drawn without lifting your pencil from the paper). In the language of Exercises 30 and 31, a unicursal diagram is

---

[14] It was Leonhard Euler who first demonstrated the impossibility of the desired scenic tour of Königsberg.

the equivalent of a plane graph with a closed Euler tour—one that ends at the same vertex from which it began. (A closed Euler tour is the edge analog of a hamiltonian cycle.) Say that a graph is *eulerian* (or *unicursal*) if it has a closed Euler tour.

(a) Prove that a connected graph on $n \geq 2$ vertices is eulerian if and only if none of its vertices has odd degree.

(b) For some random graph having 100 vertices and roughly 5000 edges, would you rather be asked to determine (1) whether it is hamiltonian or (2) whether it is eulerian? Explain.

33 A θ-graph is a 2-connected graph with two vertices of degree 3 and the rest of degree 2.

(a) Show that every θ-graph has a subgraph isomorphic to $K_{1,3}$.

(b) Show that every 2-connected nonhamiltonian graph has a θ-subgraph.

(c) Illustrate a hamiltonian graph with an induced θ-subgraph.

6
Matchings

The structure of benzene posed some interesting challenges for nineteenth century organic chemists. The same valence theory that explained Mendeleev's[1] periodic table of the elements made short work of deciphering of the structure of such compounds as hexane, chemical formula C_6H_{14}. Not only can 6 carbon atoms (valence 4) and 14 hydrogen atoms (valence 1) be made to fit together, there are five different ways to do it! For example,

$$
\begin{array}{c}
HHHHHH \\
|||||| \\
H-C-C-C-C-C-C-H \\
|||||| \\
HHHHHH
\end{array}
\quad\text{and}\quad
\begin{array}{c}
H \\
H|H \\
\backslash|/ \\
HHCH \\
|||| \\
H-C-C-C-C-H \\
|||| \\
HHCH \\
/|\backslash \\
H|H \\
H
\end{array}
$$

Valence theory did not, however, seem able to explain the existence, much less the extraordinary stability, of C_6H_6.

Assuming each carbon atom is bonded to exactly one hydrogen, it suffices to consider the structure of the underlying carbon skeleton. Initially, two competing models were proposed. In one, the carbon atoms were positioned at the six vertices of a regular octahedron. In this model, each of the 12 edges of the octahedron represents a carbon–carbon (C–C) bond. (See the plane projection of the octahedron illustrated in Fig. 6.1.) Because the valence of carbon is four, such an arrangement has the obvious defect that it does not provide for any carbon–hydrogen bonds.

The second suggestion arranged the carbon atoms at the vertices of a hexagon. In this second model, each carbon atom is bonded to its two neighbors, providing places for *two* hydrogen atoms to bond with each atom of carbon. Indeed, when benzene is induced to react with hydrogen,

$$C_6H_6 + 3H_2 \longrightarrow C_6H_{12},$$

[1] Dmitri Ivanovich Mendeleev (1834–1907) had worked out his periodic table by 1869.

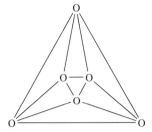

Figure 6.1. Plane projection of a regular octahedron.

the result is cyclohexane, a molecule in which the carbon atoms form a 6-membered ring, each bonded to its two neighbors, and to two hydrogen atoms.

While the octahedron model allows for too few hydrogens, the hexagon model allows for too many, *unless* there are other C—C bonds apart from those represented by the edges of the hexagon. Might the carbon skeleton of benzene exhibit one of the structures illustrated in Fig. 6.2?

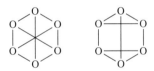

Figure 6.2

It was Baron Friedrich August Kekulé von Stadonitz (1829–1896) who solved the mystery. *Benzenediol* is the generic name for $C_6H_4(OH)_2$, a molecule that can be synthesized from benzene by replacing two hydrogen atoms with OH (oxygen–hydrogen) groups. In fact, benzenediol has three isomers: *pyrocatechol* (melting point 105°C), *resorcinal* (melting point 110°C), and *hydroquinone* (melting point 171°C). Indeed, *dichlorobenzene*, $C_6H_4Cl_2$, *dinitrobenzene*, $C_6H_4(NO_2)_2$, and a host of other compounds produced by substituting for two of the hydrogen atoms in benzene invariably come in families of three. How might a structural model account for these triples of chemical isomers?

In the benzene molecule, two carbon atoms are either bonded or not. In a regular hexagon, there is only one way for two vertices to be adjacent, but there are two distinct ways for vertices to be nonadjacent. Either they can be a distance 2 apart or they can be opposite vertices, a distance 3 apart. In the (simple) hexagon model, there are three distinct ways to replace two hydrogen atoms, depending on the graph-theoretical distance between the carbon atoms to which they are bonded. In the octahedron model, there is only one way for two vertices to be nonadjacent — they must be opposite. Because any pair of opposite vertices can be rotated into any other pair, and the same is true for any pair of adjacent vertices, the octahedron model affords just two distinct ways to replace two hydrogen atoms. Therefore, of the two models, only the hexagon can account for the experimentally observed isomers.

We are still left with the valence problem. For the atomic valences to balance in the hexagon model, there must exist some carbon–carbon bonds not accounted for by the edges of the hexagon. Kekulé rejected models like the ones in Fig. 6.2, choosing instead a solution in which single (C–C) and *double* (C=C) carbon–carbon bonds alternate around the hexagonal ring. (The two ways of doing this are illustrated in Fig. 6.3.)

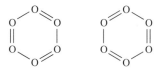

Figure 6.3

While it solves the valence problem, Kekulé's scheme invalidates the reason for choosing the hexagonal model in the first place! In Kekulé's solution, there are *two* distinct ways for a pair of carbon atoms to be adjacent! Instead of three chemical isomers, benzenediol ought to have four — *unless* the two configurations illustrated in Fig. 6.3 somehow coexist.

Experimental evidence confirms that benzene is a planar ring with all six carbon atoms in equivalent positions (and all six hydrogen atoms in equivalent positions). One (twentieth-century) theory is that the extra electrons represented by the double bonds in the Kekulé model are shared by the ring as a whole; that is, they occupy so-called "molecular orbitals." Another theory is that a rapid oscillation occurs between the two configurations of Fig. 6.3 (in each molecule of benzene). Such oscillations might account for a phenomenon called *resonance*, to which some chemists attribute the extraordinary stability of benzene.

6.1 Definition. Let $G = (V, E)$ be a graph. A *matching* of G is a subset $M \subset E$, no two edges of which are adjacent. If $e = uv \in M$, then u and v are *matched* vertices, *covered* by M. If $o(M) = r$, then M is an *r-matching*. The *matching number*, $\mu(G)$, is the largest value of r for any r-matching M of G.

Recall that a subset of $V(G)$ is independent if no two of its vertices share an edge. A matching is the analog for subsets of $E(G)$. A 1-matching is a set consisting of a single edge, matching two vertices. A 2-matching is a set of two nonadjacent edges, covering four vertices. Because an r-matching covers $2r$ vertices, $2\mu(G) \leq n = o(V(G))$.

6.2 Definition. Let G be a graph on n vertices. A *perfect* matching[2] is an r-matching for which $2r = n$.

6.3 Example. Suppose $G = C_6$. Then G has a total of 17 matchings. With respect to the numbering of its edges in Fig. 6.4, the six 1-matchings of G are $\{1\}, \{2\}, \ldots, \{6\}$. There are nine 2-matchings, namely, $\{1, 3\}, \{1, 4\}, \{1, 5\}, \{2, 4\}$, $\{2, 5\}, \{2, 6\}, \{3, 5\}, \{3, 6\}$, and $\{4, 6\}$. The two (perfect) 3-matchings, $\{1, 3, 5\}$ and $\{2, 4, 6\}$, are illustrated in Fig. 6.3. In particular, $\mu(C_6) = 3$.

Figure 6.4

While K_4 also has six 1-matchings, it has only the three 2-matchings illustrated in Fig. 6.5. Because a 2-matching in a graph with four vertices is necessarily a perfect matching, K_4 can have no 3-matchings; that is, $\mu(K_4) = 2$. □

Figure 6.5. The perfect matchings of K_4.

6.4 Definition. Let G be a graph on n vertices. Let $m(G, r)$ be the number of r-matchings of G, $1 \le r \le \mu(G)$, and set $m(G, 0) = 1$. The *matching polynomial*[3] of G is

$$M(G, x) = \sum_{r \ge 0} (-1)^r m(G, r) x^{n-2r}.$$

6.5 Example. From Example 6.3, $M(C_6, x) = x^6 - 6x^4 + 9x^2 - 2$ and $M(K_4, x) = x^4 - 6x^2 + 3$. □

---

[2] Perfect matchings are known in the chemistry literature as *Kekulé structures*.
[3] The matching polynomial (a.k.a the *acyclic* polynomial) was introduced independently by H. Hosoya in a paper on chemical thermodynamics [*Bull. Chem. Soc. Japan* **44** (1971), 2332–2339] and by O. J. Heilmann and E. H. Lieb in an article on statistical mechanics [*Commun. Math. Phys.* **25** (1972), 190–232.]

Chap. 6 Matchings

For a graph G with n vertices and m edges,

$$M(G, x) = x^n - mx^{n-2} + \cdots. \tag{39}$$

Whereas its colorful counterpart, the chromatic polynomial, was discovered more or less serendipitously, $M(G, x)$ was deliberately *contrived* to be a polynomial invariant of G. That $M(G, x)$ is a monic polynomial of degree n, with $-m$ as its next (nonzero) coefficient, makes the two polynomials seem similar. Note, however, that $-m$ is the coefficient of x^{n-2} in $M(G, x)$ and the coefficient of x^{n-1} in $p(G, x)$. Moreover, unlike the chromatic polynomial, where $p(G, r)$ is a *value* of $p(G, x)$, (plus or minus) $m(G, r)$ is a *coefficient* of $M(G, x)$. Another difference is that, while x is a factor of $p(G, x)$ for all G, it is a factor of $M(G, x)$ if and only if G does *not* have a perfect matching.

The process of comparing and contrasting the two polynomials inevitably leads to the following analog of chromatic reduction.

6.6 Theorem. *Let $G = (V, E)$ be a graph with $n \geq 3$ vertices and m edges. If $e = uv \in E$, then*

$$M(G, x) = M(G - e, x) - M(G - u - v, x). \tag{40}$$

Proof. If $r > 1$, the r-matchings that contain e are in one-to-one correspondence with the $(r-1)$-matchings of $G - u - v$, of which there are $m(G - u - v, r - 1)$. The number of r-matchings of G that do not contain edge e is $m(G - e, r)$. Thus,

$$m(G, r) = m(G - e, r) + m(G - u - v, r - 1). \tag{41}$$

Because $m(G, 1) = m$, $m(G - e, 1) = m - 1$, and $m(G - u - v, 0) = 1$, Equation (41) is also valid when $r = 1$.

By definition, the coefficient of $(-1)^r x^{n-2r}$ in $M(G, x)$ is $m(G, r)$. Similarly, $m(G - e, r)$ is the coefficient of $(-1)^r x^{n-2r}$ in $M(G - e, x)$. However, because it is a polynomial of degree $n - 2$, the coefficient of $(-1)^r x^{n-2r} = -(-1)^{r-1} x^{(n-2)-2(r-1)}$ in $M(G - u - v, x)$ is not $m(G - u - v, r - 1)$, but $-m(G - u - v, r - 1)$. Thus, $m(G - u - v, r - 1)$ is the coefficient of $(-1)^r x^{n-2r}$ in $-M(G - u - v, r - 1)$. So, Equation (40) follows from Equation (41). ∎

6.7 Corollary. *Suppose $G = (V, E)$ is a graph on $n \geq 3$ vertices. Let $u \in V$ be a vertex of degree k and suppose $N_G(u) = \{w_1, w_2, \ldots, w_k\}$. Then*

$$M(G, x) = xM(G - u, x) - \sum_{i=1}^{k} M(G - u - w_i, x), \tag{42}$$

where an empty sum (corresponding to $d_G(u) = 0$) is zero.

Proof. The proof is by induction on k. If $k = 0$, then u is an isolated vertex. In that case, $m(G, r) = m(G - u, r)$, for all r, and, because such a graph can have no perfect matchings,

$$\begin{aligned} M(G, x) &= x^n - m(G, 1)x^{n-2} + m(G, 2)x^{n-4} - \cdots \\ &= x^n - m(G - u, 1)x^{n-2} + m(G - u, 2)x^{n-4} - \cdots \\ &= x(x^{n-1} - m(G - u, 1)x^{n-3} + m(G - u, 2)x^{n-5} - \cdots) \\ &= xM(G - u, x). \end{aligned}$$

If $k > 0$, let $e = uw_k$ and $H = G - e$. From Equation (40),

$$\begin{aligned} M(G, x) &= M(G - e, x) - M(G - u - w_k, x) \\ &= M(H, x) - M(H - u - w_k, x), \end{aligned} \qquad (43)$$

because $H - u - (G - e) - u = G - u$. Since $d_H(u) = k - 1$, the induction hypothesis gives

$$M(H, x) = xM(H - u, x) - \sum_{i=1}^{k-1} M(H - u - w_i, x). \qquad (44)$$

It remains to substitute Equation (44) into Equation (43) and replace $H - u$ with $G - u$. ∎

6.8 Example. Equation (39) provides enough information to write down $M(P_1, x) = x$, $M(P_2, x) = x^2 - 1$, and $M(P_3, x) = x^3 - 2x$. If $n \geq 2$ and u is a pendant vertex of P_{n+1}, then, from Equation (42), we obtain

$$M(P_{n+1}, x) = xM(P_n, x) - M(P_{n-1}, x). \qquad (45)$$

Therefore, $M(P_4, x) = x(x^3 - 2x) - (x^2 - 1) = x^4 - 3x^2 + 1$. Similarly, $M(P_5, x) = x^5 - 4x^3 + 3x$, $M(P_6, x) = x^6 - 5x^4 + 6x^2 - 1$, $M(P_7, x) = x^7 - 6x^5 + 10x^3 - 4x$, $M(P_8, x) = x^8 - 7x^6 + 15x^4 - 10x^2 + 1$, and so on. □

6.9 Example. Theorem 6.6 lends itself to the same kind of picturesque usage as chromatic reduction. If $G = C_6$, for example, Equation (40) can be expressed as

that is, $M(C_6, x) = M(P_6, x) - M(P_4, x)$. Together with Example 6.8, this identity yields

$$M(C_6, x) = (x^6 - 5x^4 + 6x^2 - 1) - (x^4 - 3x^2 + 1)$$
$$= x^6 - 6x^4 + 9x^2 - 2,$$

confirming the first part of Example 6.5. □

6.10 Example. Consider the tree T_1 from Fig. 6.6. We have from Example 6.8 that $M(P_8, x) = x^8 - 7x^6 + 15x^4 - 10x^2 + 1$. Because it has 8 vertices (and, therefore, 7 edges) we know from Equation (39) that $M(T_1, x)$ starts off the same way; that is, $M(T_1, x) = x^8 - 7x^6 + \cdots$. Is there any reason to suppose $M(T_1, x) = M(P_8, x)$? Let's compute the rest of $M(T_1, x)$ and see.

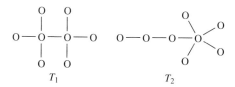

Figure 6.6

Following the approach that proved successful in Example 6.8, we might choose a pendant vertex of T_1, use Corollary 6.7 to obtain a formula for $M(T_1, x)$ similar to Equation (45), then keep repeating this procedure, eventually obtaining, as in Example 6.9, an expression involving only paths. We *might* do something hideous like that, but it should only be a last resort. Look carefully at $G = T_1$. How many 4-matchings do you see?

It is not difficult to see that T_1 does not have any 3-matchings, much less 4-matchings. Moreover, no 2-matching of T_1 can contain the "middle" edge — the one joining the two vertices of degree 4. The 2-matchings of T_1 consist of one (*pendant*) edge from the left-hand group and one form the right. By the Fundamental Counting Principle, $G = T_1$ has $3 \times 3 = 9$ 2-matchings. In other words, almost by inspection, we can write down $M(T_1, x) = x^8 - 7x^6 + 9x^4$. The easiest way to compute the matching polynomial of T_1 is directly from the definition!

While all trees on 8 vertices have the same *chromatic* polynomial, namely, $p(P_8, x) = x(x-1)^7$, it is evidently *not* the case that $M(T, x) = M(P_8, x)$ for every tree T on 8 vertices.

What about $M(T_2, x)$, where T_2 is the other tree in Fig. 6.6? Having 8 vertices it, too, starts off $M(T_2, x) = x^8 - 7x^6 + \cdots$. Take a minute to convince yourself that $m(T_2, 4) = m(T_2, 3) = 0$ and $m(T_2, 2) = 9$. Thus, $M(T_2, x) = x^8 - 7x^6 +$

$9x^4 = M(T_1, x)$. Evidently, two nonisomorphic trees on n vertices may or may not share the same matching polynomial.[4] □

6.11 Example. Consider the matching polynomial of K_n. From Equation (39), $M(K_1, x) = x$, $M(K_2, x) = x^2 - 1$, and $M(K_3, x) = x^3 - 3x$. From Example 6.5, $M(K_4, x) = x^4 - 6x^2 + 3$. If $u, w \in V(K_{n+1})$, then $d(u) = n = d(w)$, $uw \in E(K_{n+1})$, $K_{n+1} - u \cong K_n$, and $K_{n+1} - u - w \cong K_{n-1}$. So, from Corollary 6.7, we have

$$M(K_{n+1}, x) = xM(K_n, x) - nM(K_{n-1}, x), \quad n \geq 2. \tag{46}$$

In particular,

$$\begin{aligned} M(K_5, x) &= xM(K_4, x) - 4M(K_3, x) \\ &= x(x^4 - 6x^2 + 3) - 4(x^3 - 3x) \\ &= x^5 - 10x^3 + 15x. \end{aligned} \tag{47}$$

The *Hermite polynomials*[5] are defined by $h_1(x) = x$, $h_2(x) = x^2 - 1$, and $h_{n+1}(x) = xh_n(x) - nh_{n-1}(x)$, $n \geq 2$. Hermite polynomials arise as solutions to the homogeneous, second-order, linear differential equations

$$y'' - xy' + ny = 0.$$

We have just proved that $M(K_n, x) = h_n(x)$, $n \geq 1$. □

We come now to an interesting connection between the matching polynomial of G and characteristic polynomial of its adjacency matrix.

6.12 Theorem. *Let $G = (V, E)$ be a graph with vertex set $V = \{v_1, v_2, \ldots, v_n\}$. Then $M(G, x) = det(xI_n - A(G))$ if and only if G is a forest.*

Proof Sketch. $det(xI_n - A(G))$ is an alternating sum of $n!$ products, one for each permutation of $\{1, 2, \ldots, n\}$. The product corresponding to permutation p is nonzero if and only if $\{v_i, v_{p(i)}\} \in E(G)$ whenever $i \neq p(i)$, if and only if $v_i v_j \in E(G)$ whenever (ij) is a 2-cycle in the disjoint cycle factorization of p and $\langle v_i, v_j, \ldots, v_k \rangle$ is a cycle of G whenever $(ij \ldots k)$ is a cycle of p of length 3 or more. In particular, there is a one-to-one correspondence between the r-matchings

---

[4] Let $P(n)$ be the probability that a randomly chosen tree on n vertices shares its matching polynomial with a nonisomorphic tree. Then $P(n) = 0$, $n < 8$, $P(8) = 2/23$, and $P(9) = 10/47$. Any guesses for $P(10)$? (It turns out that $P(10) = 2/53$.) What about $L = \lim_{n \to \infty} P(n)$? There is, of course, no *a priori* reason for L to exist. In a remarkable *tour de force* [Almost all trees are cospectral, *New Directions in the Theory of Graphs*, Academic Press, New York, 1973, pp 275–307], A. J. Schwenk proved not only that L exists, but that $L = 1$.
[5] After Charles Hermite (1822–1901).

of G and those nonzero diagonal products of $\det(xI_n - A(G))$ afforded by permutations p whose disjoint cycle factorizations consist of r 2-cycles and $n - 2r$ fixed points (1-cycles). After attending to the signs, this one-to-one correspondence yields

$$\det(xI_n - A(G)) = M(G, x) + \text{terms involving cycles of } G.$$

If G is a forest, it doesn't have any cycles, and the proof is complete. Otherwise, one must show that there can be no cancellation among the "terms involving cycles of G." ∎

6.13 Example. It follows from Theorem 6.12 and Example 6.10 that $\det(xI_8 - A(T_1)) = x^8 - 7x^6 + 9x^4 = \det(xI_8 - A(T_2))$ for the trees T_1 and T_2 in Fig. 6.6. □

6.14 Example. Let $G = K_3$. Then G is not a forest and, as predicted by Theorem 6.12, $M(K_3, x) = x^3 - 3x \neq x^3 - 3x - 2 =$

$$\det \begin{pmatrix} x & -1 & -1 \\ -1 & x & -1 \\ -1 & -1 & x \end{pmatrix}.$$

Up to equivalence, the cycles of $G = K_3$ are $\langle v_1, v_2, v_3 \rangle$ and $\langle v_1, v_3, v_2 \rangle$, corresponding to the permutations (123) and (132), respectively. These permutations each contribute the (signed) diagonal product $(-1) \times (-1) \times (-1)$ to the characteristic polynomial of $A(G)$ and account for its constant term. □

At this point, we are going to shift our attention to the matching number and a related invariant.

6.15 Definition. Let G be a graph. A subset $B \subset V(G)$ is a *covering* of G if every edge of G is incident with a vertex of B.

If B is a covering of G, then B is said to *cover* G. So, a matching M "covers" a vertex v if some edge of the matching is incident with v, and a set B of vertices "covers" G if every edge of G is incident with some vertex of B.

If G is a graph, then $B = V(G)$ covers G. If G is bipartite with color classes X and Y, then both X and Y cover G. If the vertices of P_3 are labeled as in Fig. 6.7, then $X = \{u, w\}$ is a *minimal* covering in the sense that no proper subset of X covers P_3. On the other hand, X is not a *minimum* covering because $Y = \{v\}$ covers P_3 with fewer vertices.

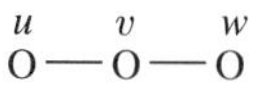

$$\overset{u}{\circ} - \overset{v}{\circ} - \overset{w}{\circ}$$

Figure 6.7

6.16 Definition. The *covering number*, $\beta(G)$, is the smallest number of vertices that cover $G = (V, E)$.

If $G = K_n^c$, then $\beta(G) = 0$. If $n \geq 2$, then $\beta(S_n) = 1$ and $\beta(K_n) = n - 1$. If $G = P_5$, then $\beta(G) = 2$. (Confirm it!)

Suppose B is a covering of $G = (V, E)$. If $e = uv \in E$, then, by definition, one (at least) of u, v is an element of B. In other words, no two vertices of $A = V \backslash B$ are adjacent. It seems that $B \subset V$ covers G if and only if $A = V \backslash B$ is an independent set. Moreover, B is a minimum covering if and only if A is a largest independent set. This observation is worth summarizing.

6.17 Lemma. *If G is a graph on n vertices, then $\alpha(G) + \beta(G) = n$.*

Suppose M is a matching of G, and B is a covering. Then every edge of M is incident with a vertex of B. Because no vertex of B can be incident with two edges of M, it must be that $o(M) \leq o(B)$. Since this inequality is independent of the matching, it must be that $\mu(G) \leq o(B)$. Finally, because this inequality is independent of the covering, it must be that

$$\mu(G) \leq \beta(G). \tag{48}$$

6.18 Egerváry–König Theorem.[6] *If $G = (V, E)$ is a bipartite graph then $\mu(G) = \beta(G)$.*

Proof. If $G \cong K_n^c$, then $\mu(G) = 0 = \beta(G)$. Otherwise, let $V = X \cup Y$ be a bipartion of G and denote by H the graph obtained from G by adding two new vertices, u and w, and n new edges, $ux, x \in X$, and $yw, y \in Y$. For any uw-path $P = [u, x_1, y_1, x_2, y_2, \ldots, x_t, y_t, w]$ in H, let $e_2(P) = x_1 y_1$. Because $\{e_2(P): P \in \mathscr{F}\}$ is a matching of G, for any family $\mathscr{F}$ of internally disjoint uw-paths in H, $\psi_H(u, w) \leq \mu(G)$. Conversely, if $M = \{x_i y_i : 1 \leq i \leq r\}$ is a matching of G, then $\{[u, x_i, y_i, w]: 1 \leq i \leq r\}$ is a family of internally disjoint uw-paths in H; that is, $\mu(G) \leq \psi_H(u, w)$. Thus, $\psi_H(u, w) = \mu(G)$. Because S separates u and w in H if and only if S is a covering of G, $\kappa_H(u, w) = \beta(G)$. Hence, the result is a consequence of Menger's Theorem. ■

Because $\mu(K_3) = 1 < 2 = \beta(K_3)$, Theorem 6.18 is not valid for graphs in general. Because $\mu(G) = \beta(G)$ for the graph G illustrated in Fig. 6.8, the converse of Theorem 6.18 is also false.

Figure 6.8. $\mu(G) = 2 = \beta(G), \chi(G) = 3$.

[6] E. Egerváry, *Mat. Lapok* **38** (1931), 16–28 and D. König, *Ibid.*, 116–119.

Pursuing these arguments a little further, suppose $G = (V, E)$ is a bipartite graph with color classes X and Y. If M is a matching of G, then $o(M) \leq o(Y)$, with equality if and only if M covers every vertex of Y, if and only if M affords a one-to-one correspondence between the elements of Y and *some* of the elements of X. Hence, $o(M) = o(Y)$ implies $o(Y) \leq o(X)$. In fact, a bit more can be said. Suppose S is a nonempty subset of Y. Let

$$N_G(S) = \bigcup_{y \in S} N_G(y)$$
$$= \{x \in X : xy \in E(G) \text{ for some } y \in S\}$$

be the set of vertices of X that are adjacent in G to some vertex of S. In order for M to cover every vertex of Y, it must, of course, cover every vertex of S, which is possible only if $o(S) \leq o(N_G(S))$.

6.19 Definition. Let $G = (V, E)$ be a bipartite graph with bipartition $V = X \cup Y$. If $o(S) \leq o(N_G(S))$, $\phi \neq S \subset Y$, then Y is said to be *expansive*.[7]

6.20 Hall's Theorem. *Let $G = (V, E)$ be a bipartite graph with bipartition $V = X \cup Y$. Then there exists a matching of G that covers every vertex of Y if and only if Y is expansive.*

Proof. The proof of necessity is what led to Definition 6.19. To prove sufficiency, suppose Y is expansive. Let M be a maximum matching of G. Because G is bipartite, every edge of M is incident with exactly one vertex of Y. Because M is a matching, no two of its edges are incident with the same vertex of Y. Together with Theorem 6.18, this implies $o(Y) \geq o(M) = \mu(G) = \beta(G)$, with equality if and only if M covers every vertex of Y. To complete the proof we will assume $o(Y) > \beta(G)$ and derive a contradiction.

Let B be minimum covering of G. If $B \subset X$, then $N_G(Y) \subset B$ and, by the hypothesis and our assumption, $\beta(G) = o(B) \geq o(N_G(Y)) \geq o(Y) > \beta(G)$, a contradiction. Thus, $Y \cap B \neq \phi$. Since $o(Y) > \beta(G) = o(B)$, $\phi \neq Y \setminus B \subset Y$. If $S = Y \setminus B$, then, because B covers G, $N_G(S) \subset X \cap B$. Therefore,

$$o(S) = o(Y) - o(Y \cap B)$$
$$= o(Y) - [o(B) - o(X \cap B)]$$
$$= o(X \cap B) + [o(Y) - \beta(G)]$$
$$> o(X \cap B)$$
$$\geq o(N_G(S)),$$

contradicting the supposition that Y is expansive. ∎

---

[7] Other names for expansive sets are *saturated* and *nondeficient*. We have chosen "expansive" because of the relation of this idea to the currently fashionable topic of *expander graphs*.

6.21 Example. Consider the bipartite graph G illustrated in Fig. 6.9. If $S = \{u, v, w\}$, then $o(N_G(S)) = 2 < o(S)$. Thus, Y is not expansive, and G does not have a perfect matching. □

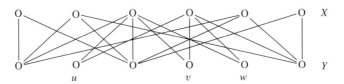

Figure 6.9

6.22 Example. Let X be a set and suppose $X_1, X_2, \ldots, X_s$ are (not necessarily different) subsets of X. In statistical design theory, $X_1, X_2, \ldots, X_s$ are sometimes called *blocks* of the *sample space* X. A *system of distinct representatives* (**SDR**) for the blocks is a collection of distinct elements $a_1, a_2, \ldots, a_s$ of the sample space X such that $a_i \in X_i$, $1 \le i \le s$.

Suppose, for example, that $X = \{1, 2, 3, 4, 5, 6\}$, $X_1 = \{1, 2, 3\}$, $X_2 = \{1, 2, 4\}$, $X_3 = \{1, 2, 5\}$, and $X_4 = \{3, 4, 5\} = X_5$. This particular sample space is small enough that the existence of an SDR is easily settled; for example, 3, 2, 1, 4, 5 is one and 1, 2, 5, 3, 4 is another.

Suppose, in general, that $X = \{x_1, x_2, \ldots, x_r\}$. Let $Y = \{X_1, X_2, \ldots, X_s\}$. Define a bipartite graph $G = (V, E)$, with vertex set $V = X \cup Y$, and with $x_i X_j \in E$ if and only if $x_i \in X_j$. Then $a_1, a_2, \ldots, a_s$ is an SDR for the blocks if and only if $M = \{a_i X_i : 1 \le i \le s\}$ is a matching of G that covers every vertex of Y. By Hall's Theorem,[8] there exists an SDR for the blocks if and only if the union of every collection of k blocks contains at least k elements, $1 \le k \le s$. □

As we saw in Fig. 6.3, the edges of C_6 can be partitioned into the disjoint union of two perfect matchings. This turns out to be a special case of a more general result.

6.23 Definition. The graph $G = (V, E)$ is *r-regular* (or *regular* of *degree r*) if $d_G(v) = r$ for all $v \in V$.

Evidently, G is regular if and only if $\delta(G) = \Delta(G)$.

6.24 Corollary. *If $G = (V, E)$ is an r-regular bipartite graph, then E is the disjoint union of r perfect matchings.*

Proof. Suppose $V = X \cup Y$ is a bipartition of V. Without loss of generality, we may assume $o(X) \le o(Y)$. Suppose $\phi \ne S \subset Y$. Then $r \times o(S)$ of the

---

[8] This is the context in which P. Hall first proved the theorem [On representatives of subsets, J. London Math. Soc. **10** (1935), 26–30].

edges of G are incident with vertices of S. These edges are all incident with the vertices of $N_G(S) \subset X$. The total number of edges incident with vertices of $N_G(S)$ is $r \times o(N_G(S))$. Therefore, $r \times o(N_G(S)) \geq r \times o(S)$. Dividing by r, we obtain $o(N_G(S)) \geq o(S)$, proving that Y is expansive. By Hall's Theorem, there is a matching M of G that covers every vertex of Y. Because $o(X) \leq o(Y)$, it must be that $o(X) = o(Y)$, and M is a perfect matching. If $r > 1$, then $H = G - M$ is an $(r-1)$-regular bipartite graph, and the result follows by induction. ∎

Let G be a (not necessarily bipartite) graph with a perfect matching M. Suppose $S \subset V(G)$. Let $H_1, H_2, \ldots, H_r$ be the components of $G - S$ that have odd numbers of vertices. Because $o(V(H_i))$ is odd, there must exist a vertex $u_i \in V(H_i)$ that is matched with a vertex $v_i \in S$, that is, $u_i v_i \in M$, $1 \leq i \leq r$. Because the u's are all different and M is a matching, the v's must all be different. Therefore, $o(S) \geq r$.

Recall that $\xi(G)$ denotes the number of (connected) components of G. Denote by $\xi_0(G)$ the number of components of G having an odd number of vertices. Using this notation, the observation in the previous paragraph may be restated as follows: If G has a perfect matching, then $o(S) \geq \xi_0(G - S)$, for all $S \subset V(G)$.[9]

6.25 Tutte's Theorem. *Let G be a graph. Then G has a perfect matching if and only if $o(S) \geq \xi_0(G - S)$ for all $S \subset V(G)$.*

Tutte's original proof of sufficiency[10] involved certain matrix functions called *pfaffians*, which are beyond the scope of this book. Anderson[11] found an elementary proof based on Hall's Theorem, and another elementary proof was discovered by Lovász.[12] While "elementary," these proofs are neither easy nor especially revealing. Thus, we will omit a proof and move on to an application.

6.26 Corollary. *If G is a 2-connected, 3-regular graph, then G has a perfect matching.*

If G were bipartite, Corollary 6.26 would be a trivial consequence of Corollary 6.24.

Proof. If G has n vertices and m edges, then, by the first theorem of graph theory, $3n = 2m$, so n is even. Since G is a connected graph on an even number of vertices, $\xi_0(G) = 0$. If G does not have a perfect matching, then, by Tutte's Theorem, there is a set $S \subset V(G)$ such that $o(S) < \xi_0(G - S)$. Moreover, $S \neq \phi$ because $o(\phi) = 0 = \xi_0(G) = \xi_0(G - \phi)$. Therefore, $1 \leq o(S) < \xi_0(G - S)$.

---

[9] This necessary condition is reminiscent of Theorem 5.3: If G has a hamiltonian cycle, then $o(S) \geq \xi(G - S)$ for all *nonempty* $S \subset V(G)$. Unlike Theorem 5.3, this necessary condition is also sufficient.
[10] W. T. Tutte, The factorization of linear graphs, *J. London Math. Soc.* **22** (1947), 107–111.
[11] I. Anderson, Perfect matchings of a graph, *J. Combinatorial Theory* B **10** (1971), 183–186.
[12] L. Lovász, Three short proofs in graph theory, *J. Combinatorial Theory* B **19** (1975), 269–271.

Consider an odd component, $H = (W, F)$ of $G - S$. Suppose $e = uv$ is an edge of G such that $u \in W$ and $v \notin W$. Since v is not in the same component of $G - S$ as u, it must be that $v \in S$. Because $\sum_{w \in W} d_H(w)$ is even and $\sum_{w \in W} d_G(w) = 3o(W)$ is odd, the number t of edges of G that join a vertex of W to a vertex of S must be odd. If $t = 1$, the unique edge joining W to S would have to be a cut-edge of G, contradicting $\varepsilon(G) \geq \kappa(G) \geq 2$. Hence, $t \geq 3$. It follows that the odd components of $G - S$ are joined to S by at least $3\xi_o(G - S) > 3o(S)$ different edges, contradicting that $\sum_{v \in S} d_G(v) = 3o(S)$. ∎

6.27 Example. The two nonisomorphic 3-regular graphs on 6 vertices are $G \cong (2K_3)^c$ and $H \cong C_6^c$. As illustrated in Fig. 6.2 (G on the left, H on the right), both of these graphs are 2-connected. Because $G \cong K_{3,3}$ is bipartite, Corollary 6.24 guarantees that $E(G)$ can be partitioned as the disjoint union of 3 perfect matchings. One way to do it is to observe that the three "diagonals" of G comprise a perfect matching, call it M. Because $G - M \cong C_6$, the two complementary perfect matchings can be found in Fig. 6.3. (Another partitioning of $E(G)$ arises from the fact that it is illustrated in Fig. 6.2 by three sets, each consisting of three parallel line segments.)

Because H is not bipartite, Corollary 6.24 does not apply to it. The weaker conclusion of Corollary 6.26 guarantees only that H has one perfect matching. In fact, $E(H)$ is also the disjoint union of 3 perfect matchings, one consisting of the three diagonals and the other two coming from C_6. This raises the natural question whether Corollary 6.26 can be strengthened. Can the edge set of every 2-connected, 3-regulator graph be partitioned into the disjoint union of 3 perfect matchings? The answer is "no." While the Petersen graph has many perfect matchings, no two of them are disjoint. (See Exercise 39, below). □

EXERCISES

1 The carbon skeletons (Hückel graphs) of naphthalene and biphenylene are illustrated in Fig. 6.10. Determine the number of perfect matchings

   (a) in napthalene.

   (b) in biphenylene.

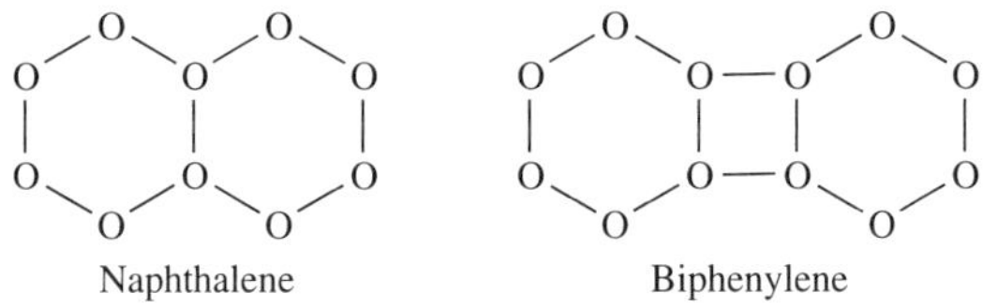

Figure 6.10

Chap. 6 Matchings

2 Compute the matching polynomial of

(a) 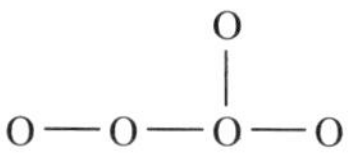 o—o—o—o   (b) 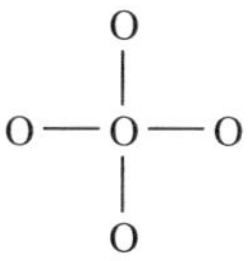 o—o(—o)(—o)(=o)   (c) 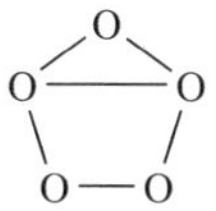 [pentagonal graph with 5 vertices]

3 According to the *resonance conjecture*, the thermodynamic stability of a "benzenoid" isomer is proportional to the number of its perfect matchings. (The more Kekulé structures, the more stable the isomer.[13]) Which of the carbon skeletons illustrated in Fig. 6.11 would the resonance conjecture predict to be the most stable? (Justify your answer.)

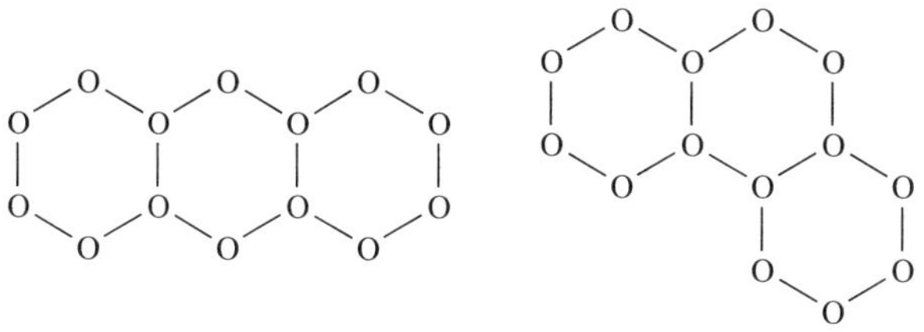

Anthracene   Phenanthrene

Figure 6.11

4 Prove that
  (a) $\mu(G)$ is an invariant.
  (b) $M(G, x)$ is an invariant.

5 Compute $M(G, x)$ for $G =$
  (a) P_9. (b) P_{10}. (c) K_6. (d) K_7.

6 Show that $M(G_1 \oplus G_2, x) = M(G_1, x) M(G_2, x)$.

7 It follows from Theorem 6.12 and properties of symmetric matrices that the roots of $M(G, x)$ are all real when G is a forest. In fact, it can be shown that the roots of $M(G, x)$ are all real for every G. Even more is true: Let $a_1 \geq a_2 \geq \cdots \geq a_n$ be the roots of $M(G, x)$.

  (a) The number of distinct roots of $M(G, x)$ is greater than the length of a longest path in G. In particular, if G has a spanning path, then the roots of $M(G, x)$ are all distinct. Confirm that $M(P_5, x)$ has distinct roots.

---

[13] The *Clar postulate* asserts that benzenoid systems without Kekulé structures are unstable biradicals. On the other hand, according to D. J. Klein et al. [Resonance in C_{60} Buckminsterfullerene, *J. Amer. Chem. Soc.* **108** (1986), 1301], the pure carbon fullerene C_{60} has 12,500 perfect matchings.

(b) If $b_1 \geq b_2 \geq \cdots \geq b_{n-1}$ are the roots of $M(G - v, x)$, then the b's *interlace* the a's; that is, $a_i \geq b_i \geq a_{i+1}$, $1 \leq i < n$. Confirm that the roots of $M(K_4, x)$ interlace the roots of $M(K_5, x)$.

8. Let c_n be the number of perfect matchings in the complete graph K_n.

   (a) Compute c_4.

   (b) Compute c_5.

   (c) Compute c_6.

   (d) Prove that $c_{n+2} = (n + 1)c_n$.

   (e) Prove that c_{2r} is odd.

9. Let G be a graph on $n \geq 3$ vertices. Suppose u and v are nonadjacent vertices of G. Then, reversing the perspective that led to Equation (40), one obtains

   $$M(G, x) = M(G + e, x) + M(G - u - v, x),$$

   where $G + e$ is the graph obtained from G by adding a new edge $e = uv$. Use this "reverse angle" approach, along with Example 6.11, to compute the matching polynomial of

   (a) $K_5 - e$.   (b) $K_6 - e$.

10. Let G be a graph on n vertices.

    (a) Prove that $\chi(G) \leq 1 + \beta(G)$.

    (b) Prove that $\chi(G) + \beta(G^c) \geq n$.

11. Let T be a tree. Prove that its adjacency matrix, $A(T)$, is invertible if and only if T has a perfect matching.

12. Find a connected graph G such that

    (a) $\alpha(G) < \beta(G)$.   (b) $\alpha(G) > \beta(G)$.

13. Let $A = (a_{ij})$ be an $m \times n$ (0, 1)-matrix (i.e., each a_{ij} is either 0 or 1). The *term rank* of A is the largest number of 1's in A, no two of which lie in the same row or column. Prove that the term rank of A is equal to the smallest cardinality of a set of rows and columns that contain every nonzero entry of A.

14. The *Hosoya topological index* of a graph G is $H(G) = \sum_{r \geq 0} m(G, r)$, the sum of the absolute values of the coefficients of the matching polynomial of G.

    (a) Prove that $H(P_n) = F(n)$, the nth *Fibonacci* number, defined recursively by $F(0) = F(1) = 1$ and $F(n + 1) = F(n) + F(n - 1)$, $n \geq 1$.

    (b) Prove that $H(C_n) = L(n)$, the nth *Lucas* number, defined by $L(0) = 2$, $L(1) = 1$, and $L(n + 1) = L(n) + L(n - 1)$, $n \geq 1$.

15 Let M be a matching of G. A path (cycle) in G is said to be M-*alternating* if the edges along the path (cycle) are alternately in M and not in M. Suppose P is an M-alternating path in G from u to w. If neither u nor w is covered by M, then P is said to be an M-*augmenting* path of G. Prove that M is a maximum matching if and only if G does not contain an M-augmenting path.

16 Let $G = (V, E)$ be a graph with matching polynomial $f(x) = M(G, x)$. It can be shown that the derivative

$$f'(x) = \sum_{v \in V} M(G - v, x).$$

(a) Use this result to prove that the Hermite polynomials satisfy $h'_n(x) = nh_{n-1}(x)$.

(b) Show how to obtain $M(K_6, x)$ by antidifferentiating Equation (47).

17 Important to the theory of matchings is the notion of adjacent edges. This concept also arises in the following way: Associated with any graph $G = (X, Y)$ is its *line graph* $G^{\#} = (Y, Z)$. The vertex set of $G^{\#}$ is $V(G^{\#}) = Y = E(G)$; that is, the vertices of $G^{\#}$ are the edges of G. The edges of $G^{\#}$ are those pairs of its vertices that are adjacent edges in G—that is, $y_1 y_2 \in Z = E(G^{\#})$, if and only if y_1 and y_2 are incident with a (single) common vertex of G.

(a) Show that the line graph of P_4 is isomorphic to P_3.

(b) Show that the line graph of K_4 is isomorphic to $K_6 - M$, where M is a perfect matching of K_6.

(c) Show that the line graph of the *wheel* W_5 = [small graph] is isomorphic to the graph in Fig. 6.12.

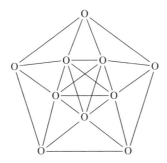

Figure 6.12

(d) Prove that $\mu(G) = \alpha(G^\#)$.

(e) Let M be a set of r edges of G. Let A be the $r \times r$ principal submatrix of $A(G^\#)$ corresponding to M. Show that M is an r-matching of G if and only if $A = 0$.

18 Let G be a bipartite graph with a spanning cycle. Prove that G has a perfect matching.

19 For each graph in Fig. 6.13, compute

(a) its degree sequence.

(b) its chromatic polynomial.

(c) its matching polynomial.

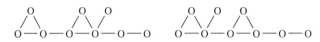

Figure 6.13

20 Using the orthogonality of the Hermite polynomials, Godsil[14] showed that

$$M(G^c, x) = \sum_{r \geq 0} m(G, r) M(K_{n-2r}, x),$$

where $M(K_0, x) = 1$. Confirm Godsil's formula for the self-complementary graph

(a) $G = P_4$. (b) $G = C_5$.

21 Say that two n-vertex graphs G and H are *neighbors* if G is isomorphic to a graph that can be obtained from H by the addition or deletion of a single edge. Then, for example, C_4 and $K_2 \oplus K_2$ are neighbors of P_4, but $K_3 \oplus K_1$ is not. Suppose V_n is a system of distinct representatives for the nonisomorphic graphs on n vertices (i.e., every graph on n vertices is isomorphic to precisely one element of V_n). Let Γ_n be the graph with vertex set $V(\Gamma_n) = V_n$, where $G, H \in V_n$ are adjacent (in Γ_n) if and only if G and H are neighbors.

(a) Prove that Γ_3 is isomorphic to P_4.

(b) Illustrate Γ_4.

(c) Prove that Γ_n is bipartite.

---

[14] C. Godsil, Hermite polynomials and a duality relation for matching polynomials, *Combinatorica* **1** (1981), 257–262.

Chap. 6 Matchings

22. Let $G = (V, E)$ be a graph on n vertices, none of which is isolated (i.e., $\delta(G) > 0$). An *edge covering* of G is a subset $L \subset E$ such that for every $v \in V$, there exists an $e \in L$ that is incident with v. The *edge covering number*, $\beta'(G)$, is the minimum number of edges in any edge covering of G.

    (a) Suppose M is a maximum matching of G. Let $U = \{u_1, u_2, \ldots, u_r\}$ be the set of vertices of G *not* covered by M. Then $r = n - 2o(M)$ is the *deficiency* of M. Show that there exist (different) edges $e_1, e_2, \ldots, e_r$ such that u_i is incident with e_i and $e_i \notin M$, $1 \le i \le r$.

    (b) Prove that $\beta'(G) \le \mu(G) + r \le n - \mu(G)$, where r is defined in part (a).

    (c) Suppose L is a minimum edge covering of G. Let $G_1 = (V(G), L)$. Show that G_1 is a graph, each of whose connected components is a star on 2 or more vertices.

    (d) Show that $\mu(G_1) = \xi(G_1) = n - \beta'(G)$, where G_1 is defined in part (c).

    (e) Show that $\mu(G) \ge n - \beta'(G)$.

    (f) Prove that $\mu(G) + \beta'(G) = n$.

    (g) If G is bipartite, prove that $\beta'(G) = \alpha(G)$.

23. In this problem, denote by S_n, not the star graph, but the set of all $n!$ permutations of $\{1, 2, \ldots, n\}$. If $A = (a_{ij})$ is an $n \times n$ matrix, its *permanent* is defined by

    $$\text{per}(A) = \sum_{p \in S_n} \prod_{t=1}^{n} a_{t p(t)}.$$

    A *Sachs* graph is a graph H, each of whose connected components is an edge or a cycle (i.e., each component of H is isomorphic to K_2 or to C_r for some r).

    (a) Let $S(G)$ be the family of *spanning* Sachs subgraphs of G. If $H \in S(G)$, let $c(H)$ be the number of components of H that are cycles. Prove that

    $$\text{per}(A(G)) = \sum_{H \in S(G)} 2^{c(H)}.$$

    (b) Denote by $K(G)$ the number of perfect matchings (Kekulé structures) in the graph G. Prove that $\text{per}(A(G)) \ge K(G)^2$.

    (c) Prove that $\text{per}(A(G)) = K(G)^2$ if G is bipartite.

24. Confirm that $\text{per}(A(G)) = K(G)^2$ (see Exercise 23) for the graph

    (a) ⧄   (b) ‖   (c) ▭   (d) ⧄

(*Hint:* Like the determinant, the permanent of an $n \times n$ matrix $A = (a_{ij})$ can be expanded by rows [or columns]. Letting A_{ij} be the submatrix of A obtained by deleting its ith row and jth column, we obtain

$$\text{per}(A) = \sum_{j=1}^{n} a_{ij}\text{per}(A_{ij}), \quad 1 \le i \le n.)$$

25 Let G be the bipartite graph illustrated in Fig. 6.9. Show that X is not expansive.

26 Let G be a graph with no isolated vertices. As in Exercise 22, an edge covering of G is a subset $L \subset E$ such that every $v \in V$ is incident with some $e \in L$. Prove the following:

  (a) A minimal edge covering of G is a minimum edge covering if and only if it contains a maximum matching.

  (b) A maximal matching is a maximum matching if and only if it is contained in a minimum edge covering.

27 Let M be a matching of G. Denote by S the set of vertices of G covered by M. Prove that there is a maximum matching of G that covers S.

28 Show that Corollary 6.26 can be strengthened as follows: Every connected, 3-regular graph with at most two cut-edges has a perfect matching. (*Hint:* In the proof of Corollary 6.26, show that o(S) and $n - $o($S$) have the same parity [both are even or both are odd] and that $n - $o($S$) = o($V(G - S)$) has the same parity as $\xi_0(G - S)$. Conclude that o(S) $\le \xi_0(G - S) - 2$.)

29 Let G be a connected graph of diameter D.

  (a) If $D \ge 3$, prove that $\mu(G) \ge \kappa(G)$.

  (b) Give an example in which $D = 2$ and $\mu(G) < \kappa(G)$.

30 Illustrate a connected, 3-regular graph that does not have a perfect matching.

31 Let $\det(xI_n - A(G)) = x^n + c_1 x^{n-1} + \cdots + c_n$ be the characteristic polynomial of the adjacency matrix of a graph G on n vertices. In 1963, H. Sachs[15] showed that

$$c_k = \sum_H (-1)^{\xi(H)} 2^{c(H)}, \quad 1 \le k \le n,$$

where the summation extends over all subgraphs H of G, on k vertices, whose connected components are either single edges or cycles, and where

---

[15] The result was obtained independently, at about the same time, by M. Milić and L. Spialter. See, e.g., D. M. Cvetković et al., *Spectra of Graphs*, 3rd. ed., Johann Ambrosius Barth Verlag, Heidelberg, 1995.

Chap. 6 Matchings

$\xi(H)$ and $c(H)$ are the numbers of components and cycles in H, respectively. Prove Sachs's Theorem.

32 Let G be a graph on n vertices. Prove that

$$M(G, x) = \sum_{M} (-1)^{o(M)} x^{\text{def}(M)},$$

where the summation is over the matchings M of G, and $\text{def}(M) = n - 2o(M)$ is the deficiency of M defined in Exercise 22(a).

33 Use Sachs's Theorem (Exercise 31) to compute the characteristic polynomial of the graph
(a) $G = P_4$. (b) $G = C_4$. (c) $G = K_4$.

34 Use Sachs's Theorem (Exercise 31) to confirm that $\det(xI_8 - A(T_1)) = x^8 - 7x^6 + 9x^4 = \det(xI_8 - A(T_2))$, where T_1 and T_2 are the trees from Fig. 6.6.

35 Let A_n be the unique, connected antiregular graph on $n \geq 2$ vertices. (See Chapter 2, Exercise 37.) Define $a(n, k) = m(A_n, k)$. The first few rows of the array $a(n, k)$ are exhibited in Fig. 6.14.

(a) Fill in row 7 of the array by finding $M(A_7, x)$.

(b) Prove that $a(n, k) = a(n-2, k) + (n - 2k + 1) a(n-2, k-1)$, for all $k \geq 1$ and $n \geq 4$.

n \ k	0	1	2	3
2	1	1		
3	1	2		
4	1	4	1	
5	1	6	4	
6	1	9	13	1

Figure 6.14. The array $a(n, k)$.

36 Prove that the cube graph Q_k of Chapter 1, Exercise 33, has a perfect matching for all $k \geq 2$.

37 The unique connected, bipartite, 2-regular graph on 8 vertices is C_8. The unique connected, bipartite, 3-regular graph on 8 vertices is the cube, Q_3.

(a) Show that the complement of Q_3 is connected and 4-regular but not bipartite.

(b) Illustrate the unique connected, bipartite, 4-regular graph on 8 vertices.

38 Prove that $\beta(G)$ is an invariant.

39 Let G be the Petersen graph in Fig. 6.15. As illustrated, G consists of an outer cycle isomorphic to C_5, an inner cycle isomorphic to $C_5^c \cong C_5$, and five "connecting" edges of which uv is one.

(a) Illustrate the two perfect matchings of G that contain edge uv.

(b) Show that every perfect matching of G contains either one or all five of the connecting edges.

(c) Show that no two perfect matchings of G are disjoint.

(d) Show that, without changing its hypothesis, the conclusion of Corollary 6.26 cannot be strenghtened to guarantee that the edge set of every 2-connected, 3-regular graph can be partitioned as the disjoint union of perfect matchings.

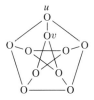

Figure 6.15. The Petersen graph.

7

Graphic Sequences

The number 6 is said to be *perfect*[1] because it is the sum of its proper divisors: $6 = 1 + 2 + 3$. Because addition is commutative and associative, this sum could just as well have been written $3 + 2 + 1$. In this context, $1 + 2 + 3$ is the same as $3 + 2 + 1$ but different, say, from $4 + 2$. In expressing the perfection of 6, what interests us is the unordered collection of its proper divisors.

7.1 Definition. Let r be a positive integer. A *partition* of r is an unordered collection of nonnegative integers that sum to r. The positive integers in the collection are the *parts* of the partition. The number of its parts, multiplicities included, is the *length* of the partition.

Among the partitions of 6 are the collections (multisets) $\{3, 3\}$ and $\{3, 3, 0\} = \{3, 0, 3\}$, both having length 2. The partitions $\{3, 1, 1, 1, 0, 0\} = \{1, 0, 1, 3, 0, 1\}$ and $\{1, 1, 1, 3\}$ both have length 4. It will greatly simplify the discussion to standardize some notation.

7.2 Definition. A partition of r is represented by a sequence $\pi = (\pi_1, \pi_2, \ldots, \pi_t)$, where $\pi_1 + \pi_2 + \cdots + \pi_t = r$ and $\pi_1 \geq \pi_2 \geq \cdots \geq \pi_t \geq 0$. This convention is expressed by means of the shorthand notation $\pi \vdash r$. If $\pi_s > 0 = \pi_{s+1}$, then $s = \ell(\pi)$, the length of π.

As used in Definition 7.2, $\pi \neq 3.1415926535\ldots$. The choice of "pi" was dictated by the fact that it shares the same first letter as "partition." At this point, it might seem more natural to write $p = (p_1, p_2, \ldots, p_t) \vdash r$. In the end, however, this choice would not result in less confusion, it would merely postpone the confusion. In fact, the potential for difficulties arising from the use of π is relatively insignificant, compared to an unfortunate semantic complication.

In ordinary English usage, arranging the collection of integers comprising a partition, from largest to smallest, would typically be referred to as "ordering" the collection. It is precisely because a partition is unordered that we are free to arrange its parts any way we like. The five cards comprising a poker hand can be arranged in any of $5! = 120$ ways, but no matter how the cards are arranged or rearranged, the hand is the same. So it is with partitions.

---

[1] A Christian theologian once explained that God, who could have created the universe in an instant, chose instead to labor for six days in order to emphasize the perfection of his creation.

Another complication involves the zeros. It might seem preferable, at least here in the beginning, to define a partition to be an unordered collection of *positive* integers. This simplification would only cause grief later on. The way to deal with zeros is to say that $\pi \vdash r$ and $\gamma \vdash r$ are *equivalent* if there is a one-to-one correspondence between the parts of π and the parts of γ in which corresponding parts are equal. Thus, π and γ are equivalent if and only if they differ, at most, in the numbers of zeros.

7.3 Example. Up to equivalence, the three-part partitions of 6 are $(4, 1, 1)$, $(3, 2, 1)$, and $(2, 2, 2)$, only one of which has anything at all to do with the perfection of 6. The equivalence class to which, for example, $(4, 1, 1)$ belongs is

$$\{(4, 1, 1), (4, 1, 1, 0), (4, 1, 1, 0, 0), (4, 1, 1, 0, 0, 0), \ldots\}. \qquad \square$$

We come, at last, to the relevance of partitions in a book about graphs.

7.4 Example. Let G be a graph with n vertices and m edges. By the first theorem of graph theory, $d(G) \vdash 2m$. The length of $d(G)$ is the number of vertices of G that have positive degree — that is, the number of nonisolated vertices of G. $\qquad \square$

We know from Chapter 1 that not every partition of $2m$ can be the degree sequence of a graph. Let's address that issue.

7.5 Definition. The partition $\pi \vdash 2m$ is *graphic* if there is a graph G whose degree sequence $d(G) = \pi$.

Suppose $\pi = (\pi_1, \pi_2, \ldots, \pi_n) \vdash 2m$. Let $\Delta = \pi_1$ and denote by π^1 the sequence obtained by rearranging (if necessary) the integers $\pi_2 - 1, \pi_3 - 1, \ldots, \pi_{\Delta+1} - 1, \pi_{\Delta+2}, \ldots, \pi_n$ into nonincreasing order. If $\pi_{\Delta+1} \geq 1$, then $\pi^1 \vdash 2(m - \Delta)$.

7.6 Theorem (Havel–Hakimi).[2] *Suppose $\pi = (\pi_1, \pi_2, \ldots, \pi_n) \vdash 2m$. Then π is graphic if and only if π^1 is graphic.*

Proof. Suppose $d(H) = \pi^1$. We may assume $V(H) = \{v_2, v_3, \ldots, v_n\}$, where $d_H(v_i) = \pi_i - 1$, $2 \leq i \leq \pi_1 + 1$. Let G be the graph obtained from H by adding a new vertex v_1, and new edges $v_1 v_i$, $2 \leq i \leq \pi_1 + 1$. Then $d(G) = \pi$, proving that the condition is sufficient.

Conversely, suppose $d(G) = \pi$, so that $\Delta = \pi_1 = \Delta(G)$. Let $u \in V(G)$ be a vertex of degree Δ. Let W be a set of Δ vertices of G having degrees $\pi_2, \pi_3, \ldots, \pi_{\Delta+1}$. If $N_G(u) = W$ (i.e., if $uw \in E(G)$ for all $w \in W$), then the graph $H = G - u$ satisfies $\pi^1 = d(H)$, and the proof is finished.

---

[2] Theorem 7.6 was published independently by V. Havel [A remark on the existence of finite graphs (Czech.), *Časopis Pěst. Mat.* **80** (1955), 477–480] and S. L. Hakimi [On the realizability of a set of integers as degrees of the vertices of a graph, *SIAM J. Appl. Math.* **10** (1962), 496–506].

Otherwise, there exist vertices $v \notin W$ and $w \in W$ such that $uv \in E(G)$ and $uw \notin E(G)$. Because $d(v) \le d(w)$, and one of the vertices adjacent to v is u, there exists a vertex $z \in V(G)$ such that $wz \in E(G)$ but $vz \notin E(G)$. This situation is illustrated in Fig. 7.1, where solid arcs join vertices known to be adjacent and dashed arcs join vertices known to be nonadjacent. (The absence of "diagonal" arcs indicates that nothing is known about the corresponding adjacencies.) Let G^1 be the graph obtained from G by *switching* the solid and dashed arcs — that is, by deleting edges uv and wz, and adding new edges vz and uw. Then $V(G^1) = V(G)$ and, because each vertex has the same degree in G^1 as in G, $d(G^1) = d(G) = \pi$. The important difference is that $\mathrm{o}(W \cap N_{G^1}(u)) > \mathrm{o}(W \cap N_G(u))$. If $N_{G^1}(u) = W$, then $H = G^1 - u$ is a graph with degree sequence π^1. Otherwise, another switch will yield a graph G^2 such that $d(G^2) = \pi$ and $\mathrm{o}(W \cap N_{G^2}(u)) > \mathrm{o}(W \cap N_{G^1}(u))$. Eventually, we obtain a graph G^k such that $\pi = d(G^k)$ and $N_{G^k}(u) = W$, at which point $H = G^k - u$ will be a graph with degree sequence $d(H) = \pi^1$. ∎

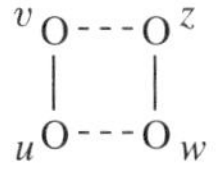

```
 v o---o z
   |   |
 u o---o w
```

Figure 7.1

7.7 Example. Suppose $\pi = (4, 4, 4, 2, 2, 1, 1) \vdash 18$. Then $\pi^1 = (3, 3, 1, 1, 1, 1)$ $\vdash 10$. By Theorem 7.6, π is graphic if and only if π^1 is graphic. Applying the theorem again, we see that π is graphic if and only if π^2 is graphic, where $\pi^2 = (2, 1, 1, 0, 0)$ is the partition of 4 obtained by rearranging the terms of the sequence 2, 0, 0, 1, 1. Because, $\pi^3 = (0, 0, 0, 0) = d(K_4^c)$, we conclude that $\pi^2 = (2, 1, 1, 0, 0), \pi^1 = (3, 3, 1, 1, 1, 1)$, and $\pi = (4, 4, 4, 2, 2, 1, 1)$ are all graphic.

In general, this reduction need not continue all the way to the bitter end. If, for example, $\pi^i = (2, 0, 0, 0, 0)$ for some i, we can stop immediately. No graph can have just one vertex of positive degree. If a negative integer appears, we can stop; no graph can have a vertex of negative degree. Had we noticed that $d(P_3 \oplus K_2^c) = (2, 1, 1, 0, 0) = \pi^2$, we could have concluded that $\pi = (4, 4, 4, 2, 2, 1, 1)$ is graphic without taking the last step.

Speaking of the last step, a review of the proof of Theorem 7.6 shows that π is graphic if and only if there is a nonnegative integer j such that π^j is a sequence of $r(= n - j)$ zeros. The number $r = r(\pi)$ of zeros that remain, when the reduction of a graphic sequence π is carried out to the bitter end, is called the Havel–Hakimi *residue* of π. At the risk of confusing it with the number of regions of a plane graph, it seems natural to write $r(G) = r(d(G))$.[3]

---

[3] It was proved by O. Favaron, M. Mahéo, and J.-F. Saclé [On the residue of a graph, *J. Graph Theory* **15** (1991), 39–64] that $r(G) \le \alpha(G)$; in other words, the residue of a graphic sequence π is a lower bound for the independence number of any graph whose degree sequence is π. (See [J. R. Griggs and D. J. Kleitman, Independence and the Havel–Hakimi residue, *Discrete Math.* **127** (1994), 209–212] for a short, revealing proof.)

If π is graphic then, by retracing the steps that produce its Havel–Hakimi residue, we can construct a graph G having degree sequence $d(G) = \pi$. Consider, for example, $\pi = (4, 4, 4, 2, 2, 1, 1)$. Earlier in this example, we found that $\pi^3 = (0, 0, 0, 0)$. Begin with $r(\pi) = 4$ isolated vertices. (Draw four dots on a sheet of paper.) To obtain a graph with degree sequence $\pi^2 = (2, 1, 1, 0, 0)$, add a new vertex w and new edges joining w to two of the existing vertices. The result is a graph isomorphic to $P_3 \oplus K_2^c$.

Before taking the next step, it is necessary to recall that π^2 arose as a rearrangement of the sequence $2, 0, 0, 1, 1$. Let H be the graph obtained from $P_3 \oplus K_2^c$ by adding a new vertex v and $d_H(v) = 3$ new edges, one joining v to w, the existing vertex of degree 2, and the others joining v to the two (remaining) isolated vertices. Then, $d(H) = \pi^1 = (3, 3, 1, 1, 1, 1)$. (Graph H is illustrated in Fig. 7.2.)

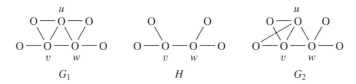

Figure 7.2

A graph G, with $d(G) = \pi = (4, 4, 4, 2, 2, 1, 1)$, can be obtained from H by adding a new vertex u and $d_G(u) = 4$ new edges. Two of these join u to the existing vertices of degree three (v and w). The other two new edges are meant to join u to two of the four vertices of degree 1. But, which two? If we join u to the pendant neighbors of v, we obtain graph G_2 in Fig. 7.2. A graph isomorphic to G_2 results from joining u to the two pendant neighbors of w. However, if we join u to one pendant neighbor of v and one pendant neighbor of w, we obtain the nonisomorphic graph G_1. Observe that $d(G_1) = \pi = d(G_2)$. □

7.8 Example. Suppose $\pi = (2, 2, 1, 1, 1, 1)$. Upon rearranging $1, 0, 1, 1, 1$ we obtain $\pi^1 = (1, 1, 1, 1, 0)$. Rearranging $0, 1, 1, 0$ yields $\pi^2 = (1, 1, 0, 0)$. One more step produces $\pi^3 = (0, 0, 0) = d(K_3^c)$. By the Havel–Hakimi Theorem, there exists a graph G having degree sequence $d(G) = \pi$. To find such a graph, begin with three isolated vertices. Adding a vertex w, along with an edge joining w to one of the existing vertices, produces $K_2 \oplus K_2^c$, a graph with degree sequence π^2. Recalling that π^2 arose as a rearrangement of $0, 1, 1, 0$, let H be the graph obtained from $K_2 \oplus K_2^c$ by adding a new vertex v and a new edge joining v to one of the isolated vertices. Then, $H \cong K_2 \oplus K_2 \oplus K_1$ and $d(H) = \pi^1$. Finally, because π^1 originated from a rearrangement of $1, 0, 1, 1, 1$, we obtain G by adding a new vertex u and two new edges, one joining u to the remaining isolated vertex and the other joining u to one of the vertices of degree 1. There is no ambiguity about G this time. No matter which vertex of degree 1 is joined to u by the last edge, the result is a graph isomorphic to $P_4 \oplus K_2$. (Confirm it!) Does this mean $P_4 \oplus K_2$ is the only graph having degree sequence $\pi = (2, 2, 1, 1, 1, 1)$? No, but it *does* mean that $P_3 \oplus P_3$ will never be produced by retracing the steps that lead to its Havel–Hakimi residue. □

Let us return to the "switching" of edges that was so useful in proving Theorem 7.6.

7.9 Definition. Let $G = (V, E)$ be a graph. Suppose $u, v, x, y \in V$ are four (different) vertices such that $uv, xy \in E$ and $ux, vy \notin E$. A *Ryser switch*[4] in G is accomplished by deleting edges uv and xy and adding edges ux and vy.

A Ryser switch in $G = (V, E)$ produces a new graph $H = (V, F)$, where $F = (E \setminus \{uv, xy\}) \cup \{ux, vy\}$. The conditions permitting a Ryser switch are illustrated (again) in Fig. 7.3, where solid arcs join adjacent vertices and dashed arcs join nonadjacent vertices. Observe that the adjacency or nonadjacency of uy and/or vx is irrelevant to a Ryser switch. What's required is that the induced subgraph, $G[u, v, x, y]$, be isomorphic to one of $2K_2, P_4$, or C_4.

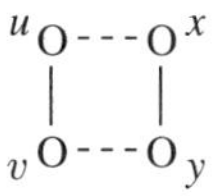

Figure 7.3

7.10 Theorem. *Let G and H be graphs. Then $d(G) = d(H)$ if and only if G can be transformed into a graph isomorphic to H by a (possibly empty) sequence of Ryser switches.*

Proof. Because $d(G_1) = d(G)$, for any graph G_1 obtained from G by a Ryser switch, sufficiency is clear.
Conversely, suppose $d(G) = d(H) = (\Delta = \pi_1, \pi_2, \ldots, \pi_n)$. If $n \leq 4$ then (see, e.g., Fig. 1.9) $G \cong H$. So, assume $n \geq 5$. Without loss of generality, we may also assume that $G = (V, E), H = (V, F)$, and $d_G(v) = d_H(v), v \in V$. Let $u \in V$ be a vertex of degree Δ (in both G and H). Let $W \subset V$ be a set of Δ vertices of degrees $\pi_2, \pi_3, \ldots, \pi_{\Delta+1}$. As we saw in the proof of Theorem 7.6, G can be transformed by

---

[4] Named for Herbert John Ryser, who introduced switching in the context of (0,1)-matrices (See, e.g., H. J. Ryser, *Combinatorial Mathematics*, Carus Mathematical Monographs, Vol. 14, Mathematical Association of America, Washington, D. C., 1963, Theorem 3.1.)

a sequence of Ryser switches into a graph G^k such that $d_{G^k}(v) = d_G(v)$, $v \in V$, and $N_{G^k}(u) = W$. Similarly, H can be transformed by a (possibly different) sequence of switches into a graph H^j such that $d_{H^j}(v) = d_H(v)$, $v \in V$, and $N_{H^j}(u) = W$.

Let $G' = G^k - u$ and $H' = H^j - u$. By induction, G' can be transformed into H' by a sequence of Ryser switches. Because $N_{G^k}(u) = N_{H^j}(u)$, and none of these switches involve u, the same sequence transforms G^k into H^j. Therefore, G can be transformed into H by the sequence of Ryser switches that changes G into G^k, followed by the sequence that changes G^k into H^j, succeeded by reversing the sequence of switches that transforms H into H^j. ∎

7.11 Corollary. *Suppose the graph G does not have an induced subgraph isomorphic to one of the three forbidden subgraphs $2K_2$, P_4, or C_4. Then G is uniquely determined by its degree sequence; that is, if $d(H) = d(G)$, then $H \cong G$.*

Proof. If $d(G) = d(H)$, then, by Theorem 7.10, G can be transformed into H by a sequence of Ryser switches. However, a Ryser switch can occur in G only if it has an induced subgraph isomorphic to one of $2K_2$, P_4, or C_4. ∎

7.12 Example. Let $G \cong 3K_2$ be the graph in Fig. 7.4(a). Because $d(v) = 1$ for all $v \in V(G)$, G is said to be *1-regular*, or *regular* of degree 1. Indeed, it is easy to see that, up to isomorphism, G is the unique 1-regular graph on 6 vertices. (Note that the graph in Fig. 7.4(b), obtained from G by a single Ryser switch, is isomorphic to G.) The complement of G is the so-called *cocktail party graph*, H, illustrated in Fig. 7.4(c). If there were another 4-regular graph on 6 vertices, its complement would be 1-regular and, therefore, isomorphic to G. Thus, despite the fact that it has induced subgraphs isomorphic to C_4, H must be the only 4-regular graph on 6 vertices. □

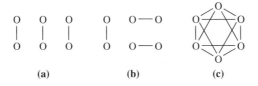

Figure 7.4

Up to equivalence, the partitions of 6 having two parts are (5, 1), (4, 2), and (3, 3). By definition, the sums of the parts of each of these partitions is 6. Their products, on the other hand, are 5, 8, and 9. This is related to the *isoperimetric* theorem of Euclid: Among all rectangles with a fixed perimeter, the one having the largest area is the square. Given that $x + y = p/2$ is fixed, $f(x, y) = xy$ attains its maximum when $x = y$. What about (9, 5, 3) and (7, 6, 4), both partitions of 17? Without doing any multiplying, is it immediately obvious that $9 \times 5 \times 3 < 7 \times 6 \times 4$? The governing principle in questions of this kind involves what has come to be known as *majorization*.

7.13 Definition. Suppose $\alpha = (a_1, a_2, \ldots, a_s)$ and $\beta = (b_1, b_2, \ldots, b_t)$ are nonincreasing sequences of real numbers. Then α *majorizes* β, written $\alpha \succ \beta$, if $s \leq t$,

$$\sum_{i=1}^{k} a_i \geq \sum_{i=1}^{k} b_i, \qquad (49)$$

$1 \leq k \leq s$, and

$$\sum_{i=1}^{s} a_i = \sum_{i=1}^{t} b_i. \qquad (50)$$

The generalization of Euclid's isoperimetric theorem is the following "product theorem": Suppose $\alpha = (a_1, a_2, \ldots, a_t)$ and $\beta = (b_1, b_2, \ldots, b_t)$ are nonincreasing sequences of length t of *positive* real numbers. If $\alpha \succ \beta$, then $\prod a_i \leq \prod b_i$.

7.14 Example. If $\pi = (9, 5, 3)$ and $\gamma = (7, 6, 4)$, then $\pi \succ \gamma$ because $9 \geq 7$, $9 + 5 \geq 7 + 6$, and $9 + 5 + 3 = 7 + 6 + 4$. Thus, by the product theorem, $9 \times 5 \times 3 \leq 7 \times 6 \times 4$. (Do the multiplications and see!) The product theorem fails if the sequences have unequal lengths or if the numbers are not all positive. While $(4, 2) \succ (3, 2, 1)$, it is not true that $4 \times 2 \leq 3 \times 2 \times 1$; and $(4, -1, -1) \succ (2, 1, 1)$, but $4 \times (-1) \times (-1) \not\leq 2 \times 1 \times 1$. □

Suppose (as in the first part of Example 7.14) that α and β are nonincreasing sequences of positive *integers*. If $\alpha \succ \beta$, then, by Equation (50), α and β are both partitions of $r = \sum a_i = \sum b_i$. In this case, majorization has a useful geometric description.

7.15 Definition. Suppose π is a t-part partition of r. The corresponding *Ferrers* (or *Young*) *diagram*[5] $F(\pi)$, consists of r "boxes," arranged in t left-justified rows. The number of boxes in row i of $F(\pi)$ is π_i, $1 \leq i \leq t$.

7.16 Example. If π and μ are equivalent partitions (differing at most in their numbers of zeros), then $F(\pi) = F(\mu)$. The left-hand side of Fig. 7.5 illustrates this situation when $\pi = (9, 5, 3)$ and $\mu = (9, 5, 3, 0)$. The fact that partitions are

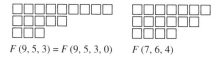

$F(9, 5, 3) = F(9, 5, 3, 0) \qquad F(7, 6, 4)$

Figure 7.5

---
[5] Named for Norman Macleod Ferrers (or Alfred Young).

represented by nonincreasing sequences means that the $(i+1)$st row of a Ferrers diagram can be no longer than its ith row, $i \geq 1$. □

If $\pi \succ \gamma$ then, from the definitions, $F(\pi)$ and $F(\gamma)$ contain the same number of boxes. Almost as obvious is the following fact, stated as a lemma for later reference.

7.17 Lemma. Suppose $\pi, \gamma \vdash r$. Then $\pi \succ \gamma$ if and only if $F(\pi)$ can be obtained from $F(\gamma)$ by moving boxes *up* (to lower numbered rows), if and only if $F(\gamma)$ can be obtained from $F(\pi)$ by moving boxes *down* (to higher numbered rows).

It is clear from Fig. 7.5 that $F(9, 5, 3)$ can be obtained from $F(7, 6, 4)$ by moving the last (right-hand end) box from each of rows 2 and 3 up to the end of row 1. A little care must be taken when moving boxes in order to ensure, at each stage, that the resulting array is a legitimate Ferrers diagram. For example, the last box in row i of $F(\pi)$ can be (re)moved only if $i = \ell(\pi)$, or $i < \ell(\pi)$ and $\pi_i > \pi_{i+1}$.

Because $F(\pi)$ determines π only up to equivalence, it will do no harm to suppose, whenever it is convenient, that the jth row of $F(\pi)$ is empty whenever $j > \ell(\pi)$. A box can be moved to the end of row j only if $j = 1$, or $j > 1$ and $\pi_j < \pi_{j-1}$. In particular, a box can always be moved to the bottom of the first column or to the end of the first row.

7.18 Lemma. Suppose $\pi, \gamma \vdash r$. If π is graphic and if π majorizes γ, then γ is graphic.

Proof. Suppose $\pi = d(G)$. By adding zeros to the end of π or γ if necessary, we may assume $\pi = (\pi_1, \pi_2, \ldots, \pi_n)$ and $\gamma = (\gamma_1, \gamma_2, \ldots, \gamma_n)$. (Adding zeros to π is equivalent to adding isolated vertices to G.) Let $V(G) = \{v_1, v_2, \ldots, v_n\}$, where $d_G(v_i) = \pi_i$, $1 \leq i \leq n$.

If $\pi = \gamma$, there is nothing to prove. Otherwise, because r is the common sum of both sequences, there is a smallest integer i such that $\pi_i > \gamma_i$ and a smallest integer t such that $\pi_t < \gamma_t$. Because $\pi \succ \gamma$, $i < t$. Let j be the largest integer such that $\pi_j = \pi_i$, and s the smallest integer such that $\pi_s = \pi_t$. Since $\pi_i - 1 \geq \gamma_i \geq \gamma_t \geq \pi_t + 1$, $\pi_j \geq \pi_s + 2$. Let w be a vertex that is adjacent in G to v_j but not to v_s. Let G_1 be the graph obtained from G by removing edge wv_j and adding edge wv_s. If $\rho = d(G_1)$, then $F(\rho)$ is obtained from $F(\pi)$ by moving the last box in row j down to the end of row s. (If v_t is an isolated vertex of G, then $s = t$ and $F(\rho)$ is obtained from $F(\pi)$ by moving the last box in row j down to the end of the first column, thus creating a new row t.) By Lemma 7.17, $\pi \succ \rho$; and from the definition of majorization, $\rho \succ \gamma$. If $\rho = \gamma$, the proof is finished. Otherwise, repeat the process, eventually arriving at a graph G_k whose degree sequence is $d(G_k) = \gamma$. ∎

7.19 Example. Suppose $\pi = (2, 2, 2)$ and $\gamma = (2, 1, 1, 1, 1)$. Then $\pi \succ \gamma$ and because $d(K_3) = \pi$, it is graphic. Let's see how the proof of Lemma 7.18

produces a graph H whose degree sequence is $d(H) = \gamma$. Begin by replacing π with $(2, 2, 2, 0, 0)$ and let $G = K_3 \oplus K_2^c$. In this case, $i = 2$, $t = 4$, $j = 3$, $s = t$, v_j is any vertex of G degree 2, and v_s is either isolated vertex. The role of w is played by a vertex of K_3 different from v_j. The graph obtained from G by removing edge wv_j and adding edge wv_s is $G_1 \cong P_4 \oplus K_1$, and $\rho = (2, 2, 1, 1, 0) \succ \gamma$.

$$G = \begin{array}{c} \circ \\ / \backslash \\ \circ - \circ \end{array} \quad \circ \quad \circ \quad , \quad G_1 \cong \begin{array}{c} \circ \\ / \\ \circ - \circ - \circ \end{array} \quad \circ$$

In the second step, $i = j = 2$, $t = s = 5$, v_j is either vertex of P_4 of degree 2, and v_s is the isolated vertex of G_1. No matter which neighbor of v_j we take for w, the graph G_2 obtained from G_1 by removing edge wv_j and adding edge wv_s is isomorphic to $P_2 \oplus P_3$, and $d(G_2) = (2, 1, 1, 1, 1) = \gamma$. □

7.20 Definition. Suppose $\gamma \vdash r$. The *conjugate* of γ is the partition $\gamma^* \vdash r$ whose jth part is the number of boxes in column j of $F(\gamma)$.

It is an immediate consequence of this definition that $F(\gamma^*)$ is the transpose of $F(\gamma)$. If, for example, $\gamma = (7, 6, 4)$, then (Fig. 7.5) $\gamma^* = (3, 3, 3, 3, 2, 2, 1)$. The number of boxes in the jth column of $F(\gamma)$ is equal to the number of rows of $F(\gamma)$ that contain at least j boxes — that is, to the number of parts of γ that are not less than j. So, the jth part of γ^* is

$$\gamma_j^* = o(\{i : \gamma_i \geq j\}). \tag{51}$$

7.21 Definition. Suppose $\gamma \vdash r$. The *trace* of γ is $f(\gamma) = o(\{i : \gamma_i \geq i\})$.

Geometrically, $f(\gamma)$ is the number of boxes on the diagonal of $F(\gamma)$, so $f(\gamma) = f(\gamma^*)$. Another way to say the same thing is that $f(\gamma)$ is the size of the so-called *Durfee square,* the largest square array of boxes that will fit in the upper left-hand corner of $F(\gamma)$. If $\gamma = (7, 6, 4)$, then (Fig. 7.5) $f(\gamma) = 3$. Notice that $F(\gamma)$ is completely determined by its first $f(\gamma)$ rows and columns.

We come now to the main result of this chapter.

7.22 Theorem (Ruch–Gutman).[6] *Suppose $\pi \vdash 2m$. Then π is graphic if and only if*

$$\sum_{j=1}^{k} \pi_j^* \geq \sum_{j=1}^{k} (\pi_j + 1), \quad 1 \leq k \leq f(\pi). \tag{52}$$

Proof. To prove the inequalities are necessary, let $\pi = d(G)$, where $V(G) = \{v_1, v_2, \ldots, v_n\}$ and $d_G(v_i) = \pi_i$, $1 \leq i \leq n$.

---

[6] Theorem 7.22 was first proved in E. Ruch and I. Gutman, The branching extent of graphs, *J. Combin. Inform. System Sci.* **4** (1979), 285–295. Also see W. Hässelbarth, Die Verzweigtheit von Graphen, *Comm. in Math. & Computer Chem.* (MATCH) **16** (1984), 3–17.

Consider, for example, the graph G in Fig. 7.6 having degree sequence $\pi = (4, 3, 2, 2, 1)$. Figure 7.6(a) illustrates a variation on the theme of Ferrers diagrams in which boxes are replaced by numbers. Specifically, every box in row i is replaced with the number i. Thus, for example, the second row consists entirely of 2's, and the number of 2's in the second row is $3 = d(v_2)$.

Figure 7.6

A second variation can be obtained from the first by rearranging the numbers, but not the shape. In Fig. 7.6(b), row i contains, in increasing order, the numbers (subscripts) of the vertices of G that are adjacent to v_i. Note that, in addition to sharing the same shape, variations (a) and (b) share the same integers, with the same multiplicities. Moreover, because the numbers in each row of variation (b) are arranged in increasing order, its first *column* must contain all of the 1's. Similarly, all of the 2's are contained in the first two columns, all of the 3's in the first three columns, and so on. In general, for any graph, the first r columns of variation (b) contain all the 1's, all the 2's, ..., and all the r's.

Because no vertex is adjacent to itself, no row of variation (b) contains its own number. So, in addition to all the 1's, the first column of variation (b) contains a number larger than 1. Therefore, $\pi_1^* \geq \pi_1 + 1$. Because the $(1, 1)$-entry of variation (b) is at least 2, and since the numbers in the first row are strictly increasing, the $(1, 2)$-entry must be at least 3. Assuming $\pi_2 \geq 2$, the $(2, 2)$-entry cannot be less than 3. (No row contains its own number.) Thus, all of the 1's, all of the 2's, and at least two numbers no smaller than 3 appear in the first two columns of variation (b). This proves $\pi_1^* + \pi_2^* \geq \pi_1 + \pi_2 + 2 = (\pi_1 + 1) + (\pi_2 + 1)$. Indeed, as long as $\pi_k \geq k$ (i.e., as long as $k \leq f(\pi)$), we can use the same approach to prove that

$$\pi_1^* + \pi_2^* + \cdots + \pi_k^* \geq \pi_1 + \pi_2 + \cdots + \pi_k + k$$
$$= (\pi_1 + 1) + (\pi_2 + 1) + \cdots + (\pi_k + 1).$$

To prove sufficiency, suppose $\pi \vdash 2m$ satisfies the conditions in (52). If one or more of the inequalities is strict, then boxes in the Ferrers diagram $F(\pi)$ can be moved up (to lower-numbered rows) until the Ferrers diagram of a partition $\gamma \vdash 2m$ is obtained for which equality holds in each of the inequalities. Because $F(\gamma)$ is obtained from $F(\pi)$ by moving boxes up, γ majorizes π. By Lemma 7.18

(with π and γ interchanged), it suffices to prove that γ is graphic whenever $\gamma_j^* = \gamma_j + 1, 1 \le j \le f(\gamma)$. This is done in the following lemma. ∎

Say that graph G is *equivalent* to graph H if G is isomorphic to a graph H' that can be obtained from H by adding or deleting isolated vertices. Thus, G and H are equivalent if they are isomorphic *to within vertices of degree* 0. In particular, if G and H are equivalent, they have the same number m of edges and (provided that $m > 0$) $d(G)$ and $d(H)$ are equivalent partitions of $2m$.

7.23 Lemma. *Suppose $m \ge 1$. Let $\pi \vdash 2m$ be a partition for which $\pi_j^* = \pi_j + 1, 1 \le j \le f(\pi)$. Then, up to equivalence, there exists a unique graph G such that $d(G) = \pi$.*

Proof. The proof is by induction on $f(\pi)$. If $f(\pi) = 1$, then $\pi = [m, 1, 1, \ldots, 1]$ (with m ones). Up to equivalence, the unique graph with this degree sequence is the star S_{m+1}. If $f(\pi) > 1$, let $\mu \vdash 2(m - \pi_1)$ be the partition whose Ferrers diagram is obtained from $F(\pi)$ by deleting its first row and column (illustrated by ■'s in Fig. 7.7). Note that $f(\mu) = f(\pi) - 1$ and $\mu_j^* = \mu_j + 1, 1 \le j \le f(\mu)$. By induction there is, up to isomorphism, a unique graph H, having no isolated vertices, such that $d(H) = \mu$. Let $G = K_1 \vee (H \oplus K_t^c)$, where t is the number of parts of π equal to 1. Then $d(G) = \pi$, proving existence.

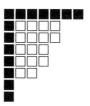

Figure 7.7

To prove uniqueness, suppose $d(G') = \pi$. From the definition of equivalence we may, without loss of generality, assume that G' has no isolated vertices. Because $\pi_1 = \pi_1^* - 1$, G' has a dominating vertex v. Moreover, $d(G' - v) = \mu$. It follows from the induction hypothesis that $G' - v$ is equivalent to H. Indeed, because G' has no isolated vertices, it must be that $G' - v$ is isomorphic to $H \oplus K_t^c$, from which it follows that G' is isomorphic to G. ∎

7.24 Definition. Suppose $m \ge 1$. If $\pi \vdash 2m$ satisfies $\pi_j^* = \pi_j + 1, 1 \le j \le f(\pi)$, it is called a *threshold partition*. If $G = K_n^c$, or if $d(G)$ is a threshold partition, then G is a *threshold graph*.[7]

---

[7] First introduced in connection with the equivalence between set packing and knapsack problems [V. Chvátal and P. L. Hammer, Aggregation of inequalities in integer programming, *Annals Disc.*

The threshold partitions are those for which equality holds throughout (52). The Ferrers diagram of the threshold partition (6, 6, 4, 3, 3, 2, 2) is exhibited in Fig. 7.8. The row of ■'s below the Durfee square accounts for the $+1$ in "$\pi_j^* = \pi_j + 1$." Notice that the part of the diagram below the row of ■'s is the transpose of the part to the right of the Durfee square.

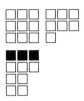

Figure 7.8

7.25 Theorem. *The partition γ is graphic if and only if it is majorized by a threshold partition.*

Proof. If γ is a graphic partition but not a threshold partition, then strict inequality holds in at least one of the inequalities in (52). As in the proof of Theorem 7.22, this means that boxes in $F(\gamma)$ can be moved up so as to produce the Ferrers diagram of a threshold partition which, by Lemma 7.17, majorizes γ. As threshold partitions are graphic (Lemma 7.23), the converse follows from Lemma 7.18. ■

7.26 Example. It follows from Theorem 7.25 that, among graphic partitions, the threshold partitions are maximal in the partial order of majorization. Figure 7.9 illustrates the partitions of 6, partially ordered by majorization. The two threshold partitions are directly below the *self-conjugate* partition (3, 2, 1). The graphic partitions of 6 are the ones whose Ferrers diagrams are constructed from dark boxes (■). □

Say that a component of G is *trivial* if it consists of an isolated vertex.

7.27 Lemma. *Suppose $G \neq K_n^c$ is a threshold graph with degree sequence $d(G) = \pi$. Let u be a vertex of G of degree π_1 and denote by C the component of G containing u. Then*

(a) C is the only nontrivial component of G;
(b) C is a threshold graph;

Math. **1** (1977), 145–162] and, independently, in the analysis of parallel behavior of systems of processes in computer programming [P. B. Henderson and Y. Zalcstein, A graph-theoretic characterization of the PV$_{\text{chunk}}$ class of synchronizing primitives, *SIAM J. Comput.* **6** (1977), 88–108], threshold graphs have been rediscovered in a variety of contexts, leading to numerous equivalent definitions.

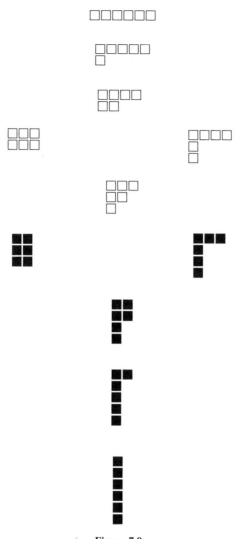

Figure 7.9

(c) u is a dominating vertex of C; and
(d) H = G − u is a threshold graph.

Proof. Observe that $\ell(\pi) = \pi_1^*$ is the number of vertices of G of positive degree. Since $\Delta = \Delta(G) = \pi_1 = \pi_1^* - 1$ it must be that, apart from itself, u is adjacent to every nonisolated vertex of G. This proves parts (a) and (c). Because $d(C)$ is equivalent to π, part (b) follows from the definitions. If $f(\pi) = 1$, then $H = G - u \cong K_{n-1}^c$ is a threshold graph. Otherwise, because $F(d(H))$ is obtained from $F(\pi)$ by deleting its first row and column, $d(H)$ is a threshold partition. ∎

One application of Lemma 7.27 is the following interesting and useful characterization of threshold graphs.

7.28 Theorem. *Let G be a graph on n vertices. Then G is a threshold graph if and only if no induced subgraph of G is isomorphic to one of the forbidden subgraphs $2K_2$, P_4, or C_4.*

Proof. There are exactly seven nonisomorphic graphs on $n \leq 3$ vertices, and it is easy to confirm that each of them is a threshold graph. Thus, the theorem is proved for $n \leq 3$, providing the basis for a proof by induction.

To prove necessity, let G be a threshold graph on $n \geq 4$ vertices. By Lemma 7.27, we may assume that G is connected, that it has a dominating vertex u, and that $H = G - u$ is a threshold graph on $n - 1$ vertices. By the induction hypothesis, H contains none of the forbidden subgraphs. Thus, if $G[W]$ is isomorphic to a forbidden subgraph, for some 4-vertex subset $W \subset V(G)$, it must be that $u \in W$. But, if $u \in W$, then either $G[W] \cong K_{1,3}$ or $G[W]$ contains a *triangle* (a subgraph isomorphic to C_3), contradicting the supposition that $G[W]$ is a forbidden subgraph.

Conversely, suppose G is a graph on $n \geq 4$ vertices that contains none of the forbidden subgraphs. Because $2K_2$ is not an induced subgraph, G has at most one nontrivial component. Hence, without loss of generality, we may assume G is connected. Let u be a vertex of G of largest degree. We claim that u is a dominating vertex of G. Otherwise, there is a vertex $x \in V(G)$ such that $ux \notin E(G)$. Because x is not an isolated vertex, there exists some $y \in V(G)$ such that $xy \in E(G)$. Because $d(u) \geq d(y)$, and because y is adjacent to a vertex not adjacent to u, there must exist a vertex $v \in V(G)$ such that $uv \in E(G)$ and $vy \notin E(G)$. But that means $G[u, v, x, y]$ is isomorphic to one of the forbidden subgraphs. Thus, u is a dominating vertex.

Let $H = G - u$. Because every induced subgraph of H is an induced subgraph of G, H contains none of the forbidden subgraphs. By induction, H is a threshold graph. So, as in the proof of Lemma 7.23 (or by Exercise 29), $H \vee \{u\} = G$ is a threshold graph. ∎

7.29 Corollary. *If G is a threshold graph, then every induced subgraph of G is a threshold graph. Moreover, G is a threshold graph if and only if G^c is a threshold graph.*

Proof. If H is an induced subgraph of G, then every induced subgraph of H is an induced subgraph of G. So, if G is a threshold graph, then no induced subgraph of H is isomorphic to a forbidden subgraph; that is, H is a threshold graph. Since $P_4^c \cong P_4$ and $C_4^c \cong 2K_2$, $G[W]$ is a forbidden subgraph of G if and only if $G^c[W]$ is a forbidden subgraph of G^c. ∎

7.30 Theorem. *Let $\pi = (\pi_1, \pi_2, \ldots, \pi_n)$, $\pi_n > 0$, be a threshold partition. Then, up to isomorphism, there is a unique threshold graph G satisfying $d(G) = \pi$. Moreover, G is connected.*

Proof. By Lemma 7.23 there is, up to isomorphism, a unique threshold graph G that has no isolated vertices and satisfies $d(G) = \pi$. By Lemma 7.27, G must be connected. ∎

Let π be a threshold partition. Because $F(\pi)$ is completely determined by its first $f(\pi)$ rows and columns, the following algorithm constructs the unique connected threshold graph corresponding to π.

7.31 Threshold Algorithm. Suppose $\pi = (\pi_1, \pi_2, \ldots, \pi_n) \vdash 2m$ is a threshold partition with $\ell(\pi) = n$.

1. Let $V = \{v_1, v_2, \ldots, v_n\}$ and $E = \phi$.
2. For $i = 1$ to $f(\pi)$.
3. For $j = i$ to π_i.
4. $E = E \cup \{v_i v_{j+1}\}$.
5. Next j.
6. Next i.
7. $G = (V, E)$ is a connected graph and $d(G) = \pi$.

Suppose, for example, that $\pi = (4, 3, 2, 2, 1) \vdash 12$, so that $m = 6$. Because $\pi_1 \geq 1$ and $\pi_2 \geq 2$, but $\pi_3 < 3$, the trace of π is $f(\pi) = 2$. Because $\pi_1^* = o(\{i: \pi_i \geq 1\}) = 5$ and $\pi_2^* = o(\{i: \pi_i \geq 2\}) = 4$, we have that $\pi_i^* = \pi_i + 1$, $1 \leq i \leq f(\pi)$; that is, π is a threshold partition. If we "run" the Threshold Algorithm with this input, the first $(i = 1)$ pass produces the graph in Fig. 7.10(a). The algorithm ends after the second $(i = 2)$ pass, and outputs the graph G illustrated in Fig. 7.10(b) (which happens to be isomorphic to the graph in Fig. 7.6). Note that $d(G) = \pi$.

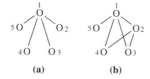

Figure 7.10

Let's examine the output of the Threshold Algorithm in general. Because $n = \pi_1^* = \pi_1 + 1$, the first $(i = 1)$ pass produces a graph with $d(v_1) = n - 1 = \pi_1$ and $d(v_2) = d(v_3) = \cdots = d(v_n) = 1$. If $f(\pi) = 1$, then $\pi_2 = 1$, and the algorithm ends having produced the star S_n. Otherwise, a second $(i = 2)$ pass *creates* edges from v_2 to v_{j+1}, $2 \leq j \leq \pi_2$. This is consistent with an input that has reserved space for $\pi_2 + 1 = \pi_2^*$ vertices of degree 2 (or more). If $f(\pi) = 2$, then, because $F(\pi)$ is completely determined by its first two rows and columns, the algorithm stops and outputs a graph having degree sequence π. Otherwise, $\pi_3 \geq 3$, and a third ($i = 3$) pass adds edges from v_3 to v_{j+1}, $3 \le j \le \pi_3$, consistent with $\pi_3 + 1 = \pi_3^*$. After three passes, the algorithm has generated a graph satisfying $d(v_1) = \pi_1$, $d(v_2) = \pi_2$, $d(v_3) = \pi_3$, and $d(v_i) \le \pi_i$, $4 \le i \le n$. After $f(\pi)$ passes, it will output a graph with $d(v_i) = \pi_i$, $1 \le i \le n$.

7.32 Example. Let $\gamma = (5, 4, 3, 3, 3, 1, 1) \vdash 20$. Then $f(\gamma) = 3$ and $\gamma^* = (7, 5, 5, 2, 1)$. Because $\gamma_1^* = 7 > 6 = \gamma_1 + 1$, $\gamma_1^* + \gamma_2^* = 12 > 11 = (\gamma_1 + 1) + (\gamma_2 + 1)$ and $\gamma_1^* + \gamma_2^* + \gamma_3^* = 17 > 15 = (\gamma_1 + 1) + (\gamma_2 + 1) + (\gamma_3 + 1)$, γ is graphic. But, it is not a threshold partition. Suppose we ignore the fact that it does not satisfy the hypothesis and input γ into the Threshold Algorithm anyway. The output (confirm it with paper and pencil) is the threshold graph illustrated in Fig. 7.11 with degree sequence $\pi = (5, 4, 3, 3, 2, 1, 0) \vdash 18$. Indeed, up to equivalence, π is the unique threshold partition that satisfies $\pi_i = \gamma_i$, $1 \le i \le f(\gamma)$. □

Figure 7.11

Recall that the vertex set of a bipartite graph can be (bi)partitioned as the disjoint union of two independent sets. As we now see, threshold graphs satisfy an analogous property.

7.33 Corollary. *Let G be a threshold graph on n vertices with $d(G) = \pi$. Let $V(G) = \{v_1, v_2, \ldots, v_n\}$, where $d_G(v_i) = \pi_i$, $1 \le i \le n$. If $W = \{v_1, v_2, \ldots, v_{f(\pi)}\}$, then $G[W]$ is a clique and $G[V \setminus W]$ is an independent set. Thus, $V(G)$ can be partitioned into two sets, one of which is independent in G and the other in G^c.*

Proof. If $i \le f(\pi)$, then $\pi_i \ge f(\pi)$ and, by the Threshold Algorithm, v_i is adjacent (at least) to $v_1, v_2, \ldots, v_{i-1}, v_{i+1}, \ldots, v_{f(\pi)}$. If $i > f(\pi)$, then v_i is adjacent only to vertices v_j for which $j \le f(\pi)$. In other words, no two of the last $n - f(\pi)$ vertices are adjacent to each other. ∎

EXERCISES

1 Exhibit (up to equivalence) all seven partitions of 5.

2 Exhibit (up to equivalence) all five 3-part partitions of 8.

3 Find two graphic partitions of 6, neither of which majorizes the other.

4 Determine whether $\pi \succ \gamma$ (and justify your answer).
   (a) $\pi = (6, 5, 4)$ and $\gamma = (5, 5, 5)$.
   (b) $\pi = (6, 5, 4)$ and $\gamma = (5, 4, 3, 2, 1)$.
   (c) $\pi = (9, 7, 5, 3)$ and $\gamma = (8, 8, 6, 2)$.

5 Find the Havel–Hakimi residue of the graphic partition $\pi =$
   (a) $(5, 4, 3, 3, 2, 1)$.   (b) $(5, 5, 5, 4, 4, 3)$.
   (c) $(6, 5, 4, 3, 3, 2, 1)$. (d) $(5, 4, 4, 4, 3, 2, 1, 1)$.

6 Exhibit a graph G having degree sequence $d(G) = \pi$ for the corresponding graphic partition of Exercise 5.

7 A partition $\pi \vdash r$ is *self-conjugate* if $\pi = \pi^*$. Find all self-conjugate partitions of 11.

8 Suppose $\pi = (2, 2, 1, 1, 1, 1) \vdash 8$. As we saw in Example 7.8, $d(G_1) = \pi = d(G_2)$, where $G_1 = P_4 \oplus K_2$ and $G_2 = P_3 \oplus P_3$.
   (a) Show that the Havel-Hakimi residue $r(\pi) \leq \alpha(G_i)$, $i = 1, 2$, where $\alpha(G)$ is the independence number of G.
   (b) Find a single Ryser switch that converts G_1 into G_2.

9 Suppose $\pi = (4, 4, 4, 2, 2, 1, 1) \vdash 18$. As we saw in Example 7.7, $d(G_1) = \pi = d(G_2)$ for the nonisomorphic graphs G_1 and G_2 of Fig. 7.2.
   (a) Show that $r(\pi) \geq \alpha(G_i)$, $i = 1, 2$.
   (b) Find a single Ryser Switch that converts G_1 into G_2.

10 Let G be the graph illustrated in Fig. 7.12(a).
    (a) Show that a Ryser switch involving the dark (solid) vertices produces the graph in either Fig. 7.12(b) or Fig. 7.12(c).

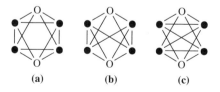

(a)   (b)   (c)

Figure 7.12

(b) Prove (by the numbers) that the graphs in Fig. 7.12 are all isomorphic (to each other).

(c) Use the notions of Example 7.12 to prove that the graphs in Fig. 7.12 are all isomorphic (to each other).

11 Let G be a graph on 4 vertices. Prove that G is isomorphic to one of $2K_2$, P_4, or C_4, if and only if
   (a) neither G nor G^c contains a triangle (a cycle of length 3).
   (b) neither G nor G^c contains a vertex of degree 3.
   (c) neither G nor G^c contains an isolated vertex.

12 Exhibit a sequence of Ryser switches that will transform the graph in Fig. 7.13(a) into the graph in Fig. 7.13(b).

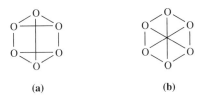

Figure 7.13

13 Prove that $\pi = (6, 5, 4, 4, 3, 2, 1, 1) \vdash 26$ is graphic
   (a) using the Havel–Hakimi Theorem.
   (b) using the Ruch–Gutman Theorem.
   (c) by finding a threshold partition that majorizes π.

14 Prove or disprove that $\pi = (8, 7, 7, 6, 6, 5, 4, 2, 1)$ is graphic
   (a) using the Havel–Hakimi Theorem.
   (b) using the Ruch–Gutman Theorem.

15 Of the 15 partitions of 7, how many are graphic?

16 As in Example 7.8, exactly one of the graphs in Fig. 7.13 can be produced by a reconstruction beginning with its Havel–Hakimi residue. Which one?

17 Prove or disprove that, up to isomorphism, there is a unique graph on 6 vertices that is regular of degree 3.

18 Show that, up to isomorphism, there is a unique connected graph having degree sequence
   (a) (3, 3, 3, 3, 2).   (b) (5, 3, 3, 2, 2, 1).
   (c) (4, 4, 4, 2, 2, 2).   (d) (3, 3, 3, 1, 1, 1).

19 Of the 22 partitions of 8,

   (a) how many are threshold?

   (b) how many are graphic?

   (c) How many nonisomorphic graphs have 4 edges and no isolated vertices?

20 Suppose r is a positive integer. Denote by $P(r)$ the set of (equivalence classes of) partitions of r. Prove that majorization is a partial order of $P(r)$. That is, prove:

   (a) $\pi \succ \pi$ for all $\pi \vdash r$.

   (b) If $\pi \succ \gamma$ and $\gamma \succ \pi$, then $\pi \equiv \gamma$.

   (c) If $\pi \succ \gamma$ and $\gamma \succ \rho$, then $\pi \succ \rho$.

21 Find two (nonisomorphic) threshold graphs each having 5 vertices and 7 edges.

22 Let A_n be the unique connected antiregular graph on n vertices. (See Chapter 2, Exercise 36.) Prove that A_n is a threshold graph.

23 Illustrate

   (a) the two connected threshold graphs with four edges.

   (b) the three connected threshold graphs with five edges.

   (c) the four connected threshold graphs with six edges.

24 (Harary and Peled) Prove that G is a threshold graph if and only if every induced subgraph of G contains an isolated vertex or a dominating vertex.

25 Denote by $t(m, n)$ the number of nonisomorphic, connected, threshold graphs with m edges and n vertices. (Then, for example $t(m, n) = 0$ if $m < n - 1$ or $m > C(n, 2)$.)

   (a) Prove that $t(m, n) = t(m - 1, n - 1) + t(m - n + 1, n - 1)$, $n - 1 \leq m \leq C(n, 2)$.

   (b) Prove that there are exactly 2^{n-2} nonisomorphic, connected, threshold graphs on $n (\geq 2)$ vertices.

26 Let G be a connected threshold graph on $n \geq 2$ vertices with degree sequence $d(G) = \pi$.

   (a) Prove that $\omega(G) = f(\pi) + 1$.

   (b) Prove that $\alpha(G) = n - f(\pi)$.

   (c) Explain why parts (a) and (b) are not inconsistent with Corollary 7.33.

27 Prove that $\gamma \succ \pi$ if and only if $\pi^* \succ \gamma^*$.

28 An *interval graph* is a graph whose vertex set is a (nonempty, finite) collection of intervals on the real line and whose edge set consists of those pairs of intervals having a nonempty intersection. The threshold graph in Fig. 7.6 is an interval graph corresponding, for example, to $v_1 = (0, 3)$, $v_2 = (0, 2)$, $v_3 = (0, 1)$, $v_4 = (1, 2)$, and $v_5 = (2, 3)$. Prove that every threshold graph is an interval graph.

29 Prove that $K_n \vee G$ is a threshold graph if and only if G is a threshold graph.

30 (Henderson and Zalcstein) A *threshold labeling* of $G = (V, E)$ is a pair (f, t), where f is an integer valued function of the vertices of G, and t is an integer (*threshold*) such that $X \subset V$ is an independent set of vertices of G if and only if $\sum_{v \in X} f(v) \leq t$.

  (a) Prove that every threshold graph has a threshold labeling.

  (b) Prove that every graph with a threshold labeling is a threshold graph.

  (c) Find a threshold labeling for the unique, connected, threshold graph G whose degree sequence is $d(G) = (5, 2, 2, 1, 1, 1)$.

31 Prove or disprove that $G \vee H$ is a threshold graph whenever G and H are connected threshold graphs.

32 (Beard and Dorris) Let G be a disconnected graph that, up to equivalence, is uniquely determined by its degree sequence π. Show that $\pi_2 \leq 1$.

33 Let G be a connected graph. The *eccentricity* of $u \in V(G)$ is $\text{ec}(u) = \max_{v \in V(G)} d(u, v)$, the maximum of the distances from u to the other vertices of G. The *radius* of G is $\text{rad}(G) = \min_{u \in V(G)} \text{ec}(u)$. The *center* of G is $\{w \in V(G) : \text{ec}(w) = \text{rad}(G)\}$. Prove that the center of a connected threshold graph is a clique.

34 (Golumbic) Let $0 < \delta_1 < \delta_2 < \cdots < \delta_k$ be the distinct degrees of the nonisolated vertices of G. Let $D_i = \{v \in V(G) : d_G(v) = \delta_i\}$. Prove that G is a threshold graph if and only if it satisfies the following condition: For all pairs of distinct vertices $u \in D_i$ and $w \in D_j$, $uw \in E(G)$ if and only if $i + j > k$.

35 Let G be a threshold graph. Suppose $uw \in E(G)$. If $d_G(u') > d_G(u)$ and $d_G(w') > d_G(w)$, prove that $u'w' \in E(G)$.

36 (Orlin) Let G be a graph that satisfies the following property: If $uw \in E(G)$, $d_G(u') > d_G(u)$, and $d_G(w') > d_G(w)$, then $u'w' \in E(G)$. Prove that G is a threshold graph.

37 Suppose $\pi \vdash 2m$ is a graphic partition. Prove that $\pi^* \succ \pi$.

38 (Gasharov) Let G be a graph with vertex set $V = \{v_1, v_2, \ldots, v_n\}$. Suppose $\{x_1, x_2, \ldots, x_n\}$ is a set of independent variables. Associate with G the monomial product

$$h(G) = \prod_{v_i v_j \in E(G)} x_i x_j,$$

and let
$$g(x_1, x_2, \ldots, x_n) = \prod_{1 \le i < j \le n} (1 + x_i x_j).$$

(a) Show that $h(G) = x_1^{d_1} x_2^{d_2} \ldots x_n^{d_n}$, where $d_i = d_G(v_i)$.

(b) Show that $h(G)$ is the term in $g(x_1, x_2, \ldots, x_n)$ corresponding to $\prod_{i<j} (x_i x_j)^{e(i,j)}$, where $e(i, j) = 1$ if $v_i v_j \in E(G)$, and 0 otherwise.

(c) If $x^\pi = x_1^{\pi_1} x_2^{\pi_2} \ldots x_n^{\pi_n}$ is among the monomials appearing in $g(x_1, x_2, \ldots, x_n)$, show that x^π is of the form $h(G)$ for some graph G having vertex set V.

(d) Show that $g(x_1, x_2, \ldots, x_n)$ is symmetric in the variables $x_1, x_2, \ldots, x_n$; that is, if cx^π occurs in $g(x_1, x_2, \ldots, x_n)$, and if μ is any (re)arrangement of π, then cx^μ is a term in $g(x_1, x_2, \ldots, x_n)$.

(e) Suppose $\pi \vdash 2m$ for some $m \le C(n, 2)$. Show that x^π is among the monomials occurring in $g(x_1, x_2, \ldots, x_n)$ if and only if π is majorized by a threshold partition of $2m$.

39 Let $G = K_5 - M$, where M is a 2-matching.

(a) Show that G is not a threshold graph.

(b) Show that if H is a graph on 5 vertices satisfying $d(H) = d(G)$ then $H \cong G$.

8

Chordal Graphs

If $\pi \vdash 2m$, then, according to the Ruch–Gutman criteria (Theorem 7.22), π is graphic if and only if

$$\sum_{j=1}^{k} \pi_j^* \geq \sum_{j=1}^{k} (\pi_j + 1), \qquad 1 \leq k \leq f(\pi). \tag{53}$$

Presented that way, the conditions seem complicated and technical. However, by extending the ideas of Chapter 7 to a slightly more general setting, we can obtain an alternative statement of the Ruch–Gutman Theorem that is both elegant and revealing. The extension is a simple one that amounts to weakening one part of the definition of majorization.

8.1 Definition. Suppose $\alpha = (a_1, a_2, \ldots, a_s)$ and $\beta = (b_1, b_2, \ldots, b_t)$ are nonincreasing sequences of real numbers. Then α *weakly majorizes*[1] β, written $\alpha \succeq_w \beta$, if $s \leq t$,

$$\sum_{i=1}^{k} a_i \geq \sum_{i=1}^{k} b_i, \qquad 1 \leq k \leq s, \tag{54}$$

and

$$\sum_{i=1}^{s} a_i \geq \sum_{i=1}^{t} b_i. \tag{55}$$

Evidently, α majorizes β if and only if α weakly majorizes β and equality holds in (55).

Suppose $\pi \vdash r$. We are going to divide $F(\pi)$ into two disjoint parts. Denote by $B(\pi)$ those boxes of $F(\pi)$ that lie strictly below its diagonal. Let $A(\pi)$ be the rest, that is, $A(\pi)$ consists of those boxes that lie on the diagonal or lie to the right of a diagonal box. Informally, $A(\pi)$ is the part of $F(\pi)$ on or *above* the diagonal, and $B(\pi)$ is the part (strictly) *below* the diagonal. If, for example,

---
[1] The standard reference for variations on the theme of majorization is A. W. Marshall and I. Olkin, *Inequalities: Theory of Majorization and Its Applications*, Academic Press, New York, 1979.

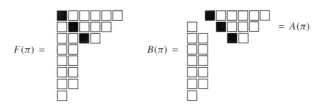

Figure 8.1

$\pi = (6, 5, 4, 2, 2, 2, 2, 1)$, the division of $F(\pi)$ into $A(\pi)$ and $B(\pi)$ is illustrated in Fig. 8.1.

8.2 Definition. Suppose $\pi \vdash r$. Let $\alpha(\pi)$ be the partition whose parts are the lengths of the rows of the *shifted shape* $A(\pi)$. Denote by $\beta(\pi)$ the partition whose parts are lengths of the *columns* of $B(\pi)$.

If $\pi = (6, 5, 4, 2, 2, 2, 2, 1)$, then (Fig. 8.1) $\alpha(\pi) = (6, 4, 2)$ and $\beta(\pi) = (7, 5)$. The number of rows of $A(\pi)$ is $\ell(\alpha(\pi)) = 3 = f(\pi)$. The jth part of $\alpha(\pi)$ is $[\alpha(\pi)]_j = \pi_j - (j - 1)$, $1 \le j \le 3$, and the jth part of $\beta(\pi)$ is $[\beta(\pi)]_j = \pi_j^* - j$, $1 \le j \le 2$. These relationships between π, π^*, $\alpha(\pi)$, and $\beta(\pi)$ are formalized in the next result.

8.3 Lemma. *Suppose $\pi \vdash r$. Let $\alpha(\pi)$ and $\beta(\pi)$ be the partitions obtained by dividing $F(\pi)$. Then*

(a) $\ell(\alpha(\pi)) = f(\pi)$;
(b) $f(\pi) - 1 \le \ell(\beta(\pi)) \le f(\pi)$;
(c) $[\alpha(\pi)]_j = \pi_j - (j - 1)$, $1 \le j \le f(\pi)$; *and*
(d) $[\beta(\pi)]_j = \pi_j^* - j$, $1 \le j \le f(\pi)$, *with the understanding that* $[\beta(\pi)]_j = 0$ *if* $j = f(\pi) > \ell(\beta(\pi))$.

Proof. Part (a) follows because each row of $A(\pi)$ begins with the corresponding diagonal box of $F(\pi)$. Part (b) is a consequence of the fact that the columns of $B(\pi)$ must fit together with the rows of $A(\pi)$ to produce a Ferrers diagram. Parts (c) and (d) are immediate consequences of the definitions. ∎

If $\pi \vdash r$, then, by (53), π is graphic if and only if

$$\sum_{j=1}^{k} \pi_j^* \ge \sum_{j=1}^{k} (\pi_j + 1), \qquad 1 \le k \le f(\pi).$$

Subtracting $k(k + 1)/2$ from both sides gives

$$\sum_{j=1}^{k} (\pi_j^* - j) \ge \sum_{j=1}^{k} (\pi_j - (j - 1)), \qquad 1 \le k \le f(\pi).$$

By Lemma 8.3, this family of inequalities is equivalent to

$$\sum_{j=1}^{k}[\beta(\pi)]_j \geq \sum_{j=1}^{k}[\alpha(\pi)]_j, \qquad 1 \leq k \leq f(\pi). \tag{56}$$

Together with Lemma 8.3(b), (56) is precisely the statement that $\beta(\pi) \succeq_w \alpha(\pi)$. This proves Roby's variation on the theme of Ruch and Gutman.[2]

8.4 Theorem. *Suppose $\pi \vdash 2m$. Then π is graphic if and only if $\beta(\pi)$ weakly majorizes $\alpha(\pi)$, that is, $\beta(\pi) \succeq_w \alpha(\pi)$.*

The family of inequalities (56) is equivalent to the family (53) because each individual inequality is equivalent, $1 \leq k \leq f(\pi)$. In particular, equality holds throughout one family of inequalities if and only if it holds throughout the other. This observation is formalized in the next result.

Theorem 8.5. *If $\pi \vdash 2m$, then π is a threshold partition if and only if $\alpha(\pi) = \beta(\pi)$.*

As we have seen, $\pi \to F(\pi)$ is a one-to-one correspondence between (equivalence classes of) partitions of $2m$ and Ferrers diagrams consisting of $2m$ boxes. It follows from Theorem 8.5 that, in this correspondence, π is a threshold partition if and only if $F(\pi)$ is "quasi-symmetric" in the sense that $A(\pi) = B(\pi)^t$, the transpose of $B(\pi)$. Because a quasi-symmetric Ferrers diagram is completely determined by its top half, $\pi \to A(\pi)$ is a one-to-one correspondence between threshold partitions of $2m$ and shifted shapes having m boxes. We know from Theorem 7.30 that there is a one-to-one correspondence between (nonisomorphic) connected threshold graphs having m edges and threshold partitions of $2m$. Putting it all together, there is a one-to-one correspondence between shifted shapes having m boxes and connected threshold graphs having m edges.[3]

Let's look a little more closely at what it means to be a "shifted shape." Unlike $F(\pi)$, the rows of $A(\pi)$ are not left-justified. Each successive row is shifted one (more) box to the right. The left-hand boundary of $A(\pi)$ looks like an inverted staircase. On the other hand, because $A(\pi)$ is just the top half of $F(\pi)$, the rules that apply to the right-hand boundary are the same for $A(\pi)$ as for $F(\pi)$, that is, the last box in row $i+1$ of $A(\pi)$ can extend no further to the right than the last box in row i. The right-hand boundary rule applied to $F(\pi)$ reflects the fact that the parts of π form a nonincreasing sequence. Because the left-hand boundary rules are different, the same right-hand rule applied to $A(\pi)$ implies that the parts of $\alpha(\pi)$ from a (strictly) decreasing sequence, that is, the parts of $\alpha(\pi)$ are all different. Conversely, if $\rho = (\rho_1, \rho_2, \ldots, \rho_k) \vdash m$ satisfies $\rho_1 > \rho_2 > \cdots > \rho_k$,

---
[2] Tom Roby is a professor at CSU Hayward.
[3] This correspondence was exploited, for example, in R. Merris and T. Roby, The lattice of threshold graphs.

then there exists a unique shifted shape whose ith row contains μ_i boxes, $1 \leq i \leq k$. It follows that the one-to-one correspondence between threshold graphs and shifted shapes extends to a one-to-one correspondence between the connected threshold graphs having m edges and the (equivalence classes of) partitions of m having distinct parts. This is interesting in part because the number of such partitions has been known at least since the time of Euler.

Corollary 8.6. *Denote by c_m the number of nonisomorphic, connected, threshold graphs having m edges. Then the generating function*

$$\sum_{m \geq 0} c_m x^m = \prod_{t \geq 1} (1 + x^t) \tag{57}$$

$$= 1 + x + x^2 + 2x^3 + 2x^4 + 3x^5 + 4x^6 \cdots.$$

Proof. Whoa! An infinite product? If you haven't seen this kind of thing before, it's not as bad as it looks. In fact, the coefficient of x^m in the right-hand side of Equation (57) is the same as the coefficient of x^m in the *finite* product

$$(1 + x) \times (1 + x^2) \times \ldots \times (1 + x^m).$$

This product is the sum of terms of the forms $x^{e_1} x^{e_2} \ldots x^{e_m}$, where the ith exponent, e_i, is either i or 0. Hence, the coefficient of x^m in the finite product is the number of ways to express $m = e_1 + e_2 + \cdots + e_m$, where $e_i = i$ or 0, that is, it is the number of (equivalence classes of) partitions of m having distinct parts. From the discussion preceding the statement of the corollary, this number is precisely c_m. ∎

By Theorem 8.4, $\pi \vdash 2m$ is graphic if and only if $\beta(\pi) \succeq_w \alpha(\pi)$. From Theorem 8.5, π is a threshold partition if and only if $\beta(\pi) = \alpha(\pi)$. This raises a more or less obvious question. Is there anything special about partitions π that satisfy $\beta(\pi) \succ \alpha(\pi)$? For one thing, $B(\pi)$ and $A(\pi)$ contain the same number of boxes. There is a positive integer m such that $\alpha(\pi) \vdash m$ and $\beta(\pi) \vdash m$. Because majorization implies weak majorization, $\beta(\pi) \succ \alpha(\pi)$ means $\pi \vdash 2m$ is graphic and m is the number of edges in any graph whose degree sequence is equivalent to π.

8.7 Definition. Suppose $\pi \vdash r$. If $\beta(\pi)$ majorizes $\alpha(\pi)$, then π is a *split* partition. If $G = K_n^c$ or if $d(G)$ is a split partition, then G is a *split* graph.

Evidently, every threshold graph is a split graph. Perhaps the simplest example of a nonthreshold split graph is P_4 whose degree sequence $d(P_4) = \gamma = (2, 2, 1, 1)$ divides as $\alpha(\gamma) = (2, 1)$ and $\beta(\gamma) = (3)$.

8.8 Example. If $\pi = (4, 4, 4, 2, 2, 1, 1)$, then (Fig. 8.2) $\beta(\pi) = (6, 3) \succ (4, 3, 2) = \alpha(\pi)$, so π is a split partition. Thus, the nonisomorphic graphs G_1 and G_2 from Example 7.7 (reproduced in Fig. 8.3) are split graphs. Evidently,

Chap. 8 Chordal Graphs

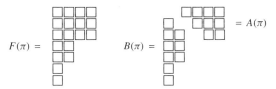

Figure 8.2

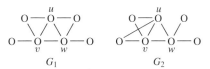

Figure 8.3

a connected split graph is not necessarily uniquely determined by its degree sequence. □

Recall (Corollary 7.33) that the vertex set of a threshold graph can be partitioned into the disjoint union of a clique and an independent set. Our proof that this property characterizes split graphs requires some preparation.

Let $\pi = (4, 4, 4, 2, 2, 1, 1)$ be the split partition from Example 8.8. Consider the unique threshold partition γ determined by $A(\gamma) = A(\pi)$. Then (Fig. 8.2), $\beta(\gamma) = \alpha(\gamma) = \alpha(\pi) = (4, 3, 2)$, so $\gamma = (4, 4, 4, 3, 3)$. (Confirm it with paper and pencil.) Because $\beta(\pi) \succ \alpha(\pi) = \beta(\gamma)$, $B(\pi)$ can be obtained from $B(\gamma)$ by moving boxes *down*. (If this seems backwards, it is because the ith part of $\beta(\pi)$ is the number of boxes in row i, not of $B(\pi)$ but of its transpose. See Fig. 8.4.) Therefore, $F(\pi)$ can be obtained from $F(\gamma)$ by moving boxes down. So, by Lemma 7.17, the split partition π is majorized by the threshold partition γ.

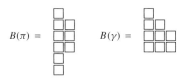

Figure 8.4

Inputting $\gamma = (4, 4, 4, 3, 3)$ into the Threshold Algorithm produces the threshold graph $T \cong K_5 - e$ shown in Fig. 8.5(a). In this case, $f(\gamma) = 3$ and, as guaranteed by Corollary 7.33, $W = \{v_1, v_2, v_3\}$ is a clique in T and $U = V \backslash W = \{v_4, v_5\}$ is an independent set.

Let's see if we can figure out what is going on *graph theoretically* as boxes are moved down in $F(\gamma)$, ultimately to produce $F(\pi)$. Because $A(\gamma) = A(\pi)$,

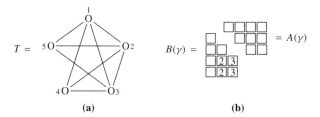

Figure 8.5

only boxes starting out in $B(\gamma)$ are eligible to be moved. The boxes in column j of $B(\gamma)$ come from column j and rows $j+1$ through $\pi_j^* = \pi_j + 1$ of $F(\gamma)$. Looking back at the second variation of $F(\gamma)$ (see Fig. 7.6(b)), there is a one-to-one correspondence between the boxes in column 1 of $B(\gamma)$ and the edges of T incident with v_1. The boxes in column 2 correspond to the edges joining v_2 to v_{i+1}, $2 \le i \le \pi_2$. In general, for $1 \le j \le f(\gamma)$, the boxes in column j of $B(\gamma)$ correspond to the edges joining v_j to v_{i+1}, $j \le i \le \pi_j$. In other words, box (i, j) of $B(\gamma)$ (the box in row $i+1$ and column $j \le i$ of $F(\gamma)$) corresponds to edge $v_{i+1}v_j$ of T. Because $E(T) = \{v_{i+1}v_j : 1 \le j \le f(\gamma), j \le i \le \pi_j\}$, this establishes a one-to-one correspondence between the boxes of $B(\gamma)$ and the edges of T. (The "other" box of $F(\gamma)$ corresponding to edge $v_{i+1}v_j$ lies in row j of $A(\gamma)$.)

Not every box in $B(\gamma)$ is eligible to be moved. No box in the first column can be moved down (and maintain a Ferrers diagram). Because the resulting configuration of boxes must fit together with $A(\gamma)$ to make a Ferrers diagram, the first (topmost) box in column j must remain in place for all $j < f(\gamma)$. The division of $F(\gamma)$ is illustrated in Fig. 8.5(b), where boxes not eligible to be moved down are empty. If box (i, j) in $B(\gamma)$, corresponding to edge $v_{i+1}v_j$ of T, is eligible to move, then, because we want to keep track of its adjacency to v_j, it is marked with (its column) number j.

In this particular example, the process of moving boxes down can only begin with box $(4,3)$ in the lower right-hand corner of $B(\gamma)$. Moreover, the only place this box can be moved to is position $(5,1)$, at the bottom of column 1. Moving box $(4,3)$ out of row 4 of $B(\gamma)$ (row 5 of $F(\gamma)$) is equivalent to detaching edge v_5v_3 from vertex v_5 (but leaving the other end firmly attached to v_3). Moving it into row 5 of $B(\gamma)$ (row 6 of $F(\gamma)$) is equivalent to reattaching the loose end to the formerly isolated vertex v_6. (As we proceed, it will be convenient to think of the independent set U as having been augmented by a sufficiently large number of isolated vertices, $v_6, v_7, \ldots$) Having modified T by removing edge v_5v_3 and adding edge v_6v_3, we arrive at the graph H in Fig. 8.6 corresponding to the (divided Ferrers) diagram in Fig. 8.6.(a). Note that $W = \{v_1, v_2, v_3\}$ is still a clique and (its augmented complement) $U = \{v_4, v_5, v_6, \ldots\}$ is still an independent set.

There are three options for moving boxes down from the configuration in Fig. 8.6(a). The bottom ② from the second column could be moved down to the end of the first column, or the remaining ③ from the third column might be moved

Chap. 8 Chordal Graphs 153

down to the end of either the first or the second column. The last of these options poses a technical difficulty. Moving box ③ from column 3 to the end of column 2 is fine, from the perspective of Ferrers diagrams. Detaching edge v_4v_3 from vertex v_4 and reattaching the loose end to v_6, on the other hand, would result in a *second* edge joining v_3 to v_6. In practice, this difficulty is easily circumvented. The boxes in row 3 correspond to the neighbors of v_4, namely, v_1 (the unlabeled box), v_2, and v_3. From a *graph* perspective, it doesn't matter whether the boxes appear in the order □②③ or □③②. For our purposes, the box diagram in Fig. 8.6(a) could just as well be replaced with the diagram in Fig. 8.6(b). Without changing any options, this replacement eliminates the technical difficulty.

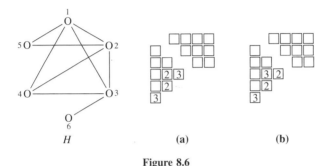

Figure 8.6

Because, in this example, we are interested in transforming $B(\gamma)$ into $B(\pi)$, let's move the last box in column 3 (labeled ③ in Fig. 8.6(a) and ② in Fig. 8.6(b)) down to the bottom of the first column. In the case of Fig. 8.6(b), moving box ② corresponds in H to detaching edge v_4v_2 from vertex v_4 and reattaching the loose end to (the formerly isolated vertex) v_7. This produces the graph G_1 in Fig. 8.7. In the case of Fig. 8.6(a), moving box ③ corresponds in H to detaching edge v_4v_3 from vertex v_4 and reattaching the loose end to vertex v_7, producing the graph G_2 in Fig. 8.7. While these (split) graphs share the degree sequence

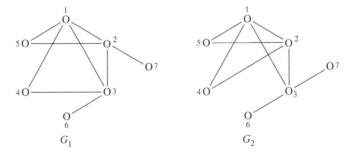

Figure 8.7

$\pi = (4, 4, 4, 2, 2, 1, 1)$, they are not isomorphic. The two pendant vertices share a common neighbor in G_2, but not in G_1. Nevertheless, $W = \{v_1, v_2, v_3\}$ is still a clique in both G_1 and G_2 and (its augmented complement) $U = \{v_4, v_5, v_6, v_7\}$ is still an independent set. Finally (and not unexpectedly), G_1 and G_2 are isomorphic, respectively, to the graphs illustrated in Fig. 8.3.

Summary. Suppose π is a split partition. Let γ be the unique threshold partition with $A(\gamma) = A(\pi)$. Inputting γ into the Threshold Algorithm, let T be the output, so that $W = \{v_i: 1 \leq i \leq f(\pi)\}$ is a clique. Augment the independent set $U = V(T) \setminus W$ with sufficiently many isolated vertices. Then each step in the process of moving boxes down in $F(\gamma)$ to produce $F(\pi)$ corresponds, graph theoretically, to deleting an edge uw, $u \in U$ and $w \in W$, and adding a new edge $u'w$, where $u' \in U$ is a higher numbered vertex than u.

8.9 Theorem. *Let G be a graph with n vertices and m edges. Then G is a split graph if and only if $V(G)$ can be partitioned into the disjoint union of an independent set and a clique (with the understanding that either set may be empty).*

Proof. If $n \leq 3$ and/or $m \leq 2$, the result follows from an easy inspection of cases. So, we may assume $n \geq 4$ and $m \geq 3$.

Suppose G is a split graph with degree sequence $d(G) = \pi$. Let γ be the unique threshold partition such that $\alpha(\gamma) = \alpha(\pi)$, and denote by T the unique connected threshold graph with degree sequence $d(T) = \gamma$. Suppose $V(T) = \{v_1, v_2, \ldots, v_t\}$ where $d_T(v_i) = \gamma_i$, $1 \leq i \leq t$. If $W = \{v_i: 1 \leq i \leq f(\pi)\}$, then, by Corollary 7.33, W is a clique and $V(T) \setminus W$ is an independent set in T. Form a vertex set U by augmenting $V(T) \setminus W$ with sufficiently many isolated vertices. As in the previous discussion, T can be modified to produce a graph G', where $d(G') = \pi$, $V(G') = U \cup W$, U is an independent set in G', and W is a clique in G'. If we could just be sure that every graph with degree sequence $\pi = d(G)$ could be obtained by such a modification of T, the proof of necessity would be finished. What we *can* be sure of is that G' can be transformed into G by a sequence of Ryser switches. (See Theorem 7.10.)

Suppose $H = (V, E)$ is any graph whose vertex set is partitioned into a clique W and independent set $U = V \setminus W$. Let $X = \{u, v, x, y\} \subset V$ be a set of four (different) vertices such that $uv, xy \in E$ and $ux, vy \notin E$. (See Fig. 8.8.) Then,

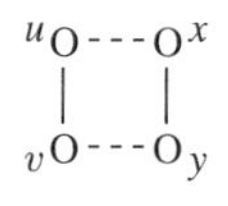

```
u o---o x
  |   |
v o---o y
```

Figure 8.8

as we have seen previously, the induced subgraph $H[X] \cong 2K_2$, P_4, or C_4. If as many as three vertices of X were to belong to W, then $H[X]$ would contain a

Chap. 8 Chordal Graphs

triangle, which it does not. If fewer than two vertices of X belong to W, then $H[X] \cong K_{1,3}$ or $H[X]$ has an isolated vertex, neither of which is correct. Thus, $o(X \cap W) = 2$. Without loss of generality, we may assume $u \in W$. If $v \in W$, then xy would be an edge joining two vertices of an independent set. Hence, $v \notin W$. If $x \in W$, then, because W is a clique, ux would be an edge of H, which it is not. So, it must be that $u, y \in W$ and $v, x \in U$. Therefore, all four of the edges involved in the corresponding Ryser switch, namely uv, xy, ux, and vy, join a vertex of W with a vertex of U. In other words, after the Ryser switch, W is still a clique and U is still an independent set. Because G can be obtained from G' by a sequence of Ryser switches, $V(G)$ can be partitioned into a clique and an independent set.

Conversely, suppose $G = (V, E)$ is a graph whose vertex set $V = \{v_1, v_2, \ldots, v_n\}$ can be partitioned into the disjoint union of a clique K and an independent set $V \backslash K$. If there is more than one such partitioning of V, choose one in which K is a maximal clique — that is, so that no vertex of $V \backslash K$ is adjacent to every vertex of K. Thus, if $s = o(K)$, then no vertex of $V \backslash K$ has degree larger than $s - 1$, and every vertex of K has degree at least $s - 1$. Let $d(G) = \pi$, where $d_G(v_i) = \pi_i$, $1 \le i \le n$. Then

$$\pi_1 \ge \pi_2 \ge \cdots \ge \pi_s \ge s - 1 \ge \pi_{s+1} \ge \cdots \ge \pi_n,$$

and $K = \{v_1, v_2, \ldots, v_s\}$.

If $\pi_s = s - 1$, then v_s is adjacent to no vertex of the independent set $V \backslash K$. In this case, let $W = \{v_i : 1 \le i < s\}$. If $\pi_s \ge s$, let $W = K$. If $o(W) = t$, then $t = s - 1$ in the first case and $t = s$ in the second. Either way, $W = \{v_1, v_2, \ldots, v_t\}$ is a clique, $V \backslash W$ is an independent set,

$$\pi_1 \ge \pi_2 \ge \cdots \ge \pi_t \ge t \ge \pi_{t+1} \ge \cdots \ge \pi_n,$$

and $f(\pi) = t$.

Let a_i be the number of vertices of $V \backslash W$ adjacent to v_i, $1 \le i \le t$. Then

$$m = a_1 + a_2 + \cdots + a_t + t(t-1)/2.$$

Because the ith row of $A(\pi)$ contains

$$[\alpha(\pi)]_i = d_G(v_i) - (i - i)$$
$$= (t - 1) + a_i - (i - 1)$$

boxes, $1 \le i \le t = f(\pi)$, $\alpha(\pi)$ is a partition of

$$\sum_{i=1}^{t} [\alpha(\pi)]_i = t(t-1) + (m - t(t-1)/2) - t(t-1)/2$$
$$= m.$$

Because $\alpha(\pi)$ is a partition of m, the number of edges of G, it must be that $\beta(\pi)$ is also a partition of m. Together with Theorem 8.4, this proves $\alpha(\pi) \succ \beta(\pi)$, that is, $\pi = d(G)$ is a split partition. ∎

8.10 Corollary. *Every induced subgraph of a split graph is a split graph.*

Proof. Let $G = (V, E)$ be a split graph, where $V = U \cup W$ is the disjoint union of an independent set U and a clique W. If $H = G[X]$ for some $X \subset V$, then $X = V(H)$ is the disjoint union of the independent set $U \cap X$ and the clique $W \cap X$. ∎

8.11 Corollary. *Let G be a graph. Then G is a split graph if and only if G^c is a split graph.*

Proof. If $V = V(G) = U \cup W$ is the disjoint union of an independent set U and a clique W, then $V(G^c) = V = W \cup U$ is the disjoint union of an independent set W and a clique U. ∎

8.12 Definition. The graph G is *chordal* if it does not contain an induced subgraph isomorphic to C_k for any $k \geq 4$.

If G is chordal and C_k is a cycle in G with $k \geq 4$ vertices, then $E(G)$ contains a *chord* of C_k—that is, an edge joining two of its (nonconsecutive) vertices.

8.13 Example. The graphs in Figs. 8.9(a) and 8.9(b) are chordal; $K_1 \vee C_6$, the *wheel* illustrated in Fig. 8.9(c), is not. Its "rim" is a chordless 6-cycle. □

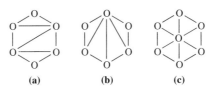

(a)   (b)   (c)

Figure 8.9

8.14 Theorem. *Suppose $G = (V, E)$ is a graph on n vertices. Then the following three conditions are equivalent.*

(a) *G is a split graph;*
(b) *G and G^c are both chordal;*
(c) *G does not contain an induced subgraph isomorphic to one of the three forbidden subgraphs $2K_2$, C_4, or C_5.*

Proof. Because the result is trivial when $n \leq 3$, we begin by assuming $n \geq 4$.

(a) $\Rightarrow$ (b): Let G be a split graph where $V = U \cup W$ is the disjoint union of an independent set U and a clique W. Suppose $C = \langle x_1, x_2, x_3, x_4, \ldots \rangle$ is a cycle of G of length $k \geq 4$. If every vertex of C belongs to W, then the path $[x_1, x_3]$ is a chord of C. Otherwise, without loss of generality, we may assume

Chap. 8 Chordal Graphs

$x_2 \in U$. Because no two consecutive vertices of C can belong to U, $x_1, x_3 \in W$ and $[x_1, x_3]$ is a chord of C. Thus, G is chordal. Because (Corollary 8.11) G^c is a split graph, the same argument proves that it is chordal.

(b) $\Rightarrow$ (c): Because G is chordal, it does not contain an induced subgraph isomorphic to C_n for any $n \geq 4$. In particular, it contains neither C_4 nor C_5. Because G^c is chordal, it does not contain an induced subgraph isomorphic to C_4. Thus, G does not contain an induced subgraph isomorphic to $C_4^c \cong 2K_2$.

(c) $\Rightarrow$ (a): Suppose G is a graph that contains none of the three forbidden subgraphs. The absence of an induced subgraph isomorphic to $2K_2$ implies that G has at most one nontrivial component. If the clique number $\omega(G) = 1$, then $G = K_n^c$ is a split graph.

Suppose $\omega(G) = 2$. Assume C is a cycle of G of minimum length r. Because $C_3 \cong K_3$, $r \geq 4$. Because no cycle of G has length less than r, C must be chordless; that is, $C \cong C_r$ is an induced subgraph of G. Because C_4 and C_5 are forbidden, $r \geq 6$, in which case C contains an induced subgraph isomorphic to $2K_2$. Therefore, G cannot have any cycles at all; that is, G is a graph whose only nontrivial component is a tree, call it T. If the diameter of T is 4 (or more), then T contains an induced subgraph isomorphic to $2K_2$. If the diameter of T is 3, then all but two of the vertices of T are pendant vertices (an independent set) and the two nonpendant vertices are adjacent (forming a clique). If the diameter of T is 2 or 1, then T is a star. Because stars are threshold graphs, they are split graphs.

Among the cliques W of G satisfying $o(W) = \omega(G) \geq 3$, choose one that minimizes the number of edges in the subgraph induced by $U = V \setminus W$. If U is an independent set, the proof is finished. Otherwise, there exist $u_1, u_2 \in U$ such that $u_1 u_2 \in E$. Because W is a maximum clique, neither u_1 nor u_2 is adjacent to every vertex of W. Indeed, it cannot even happen that they are both adjacent to every vertex of W except one, say v, because in that case $(W \setminus \{v\}) \cup \{u_1, u_2\}$ would be a clique in G having more than $\omega(G)$ vertices. Thus, there exist vertices $w_1, w_2, \in W$ such that $w_1 \neq w_2$, $u_1 w_1 \notin E$, and $u_2 w_2 \notin E$. (See Fig. 8.10(a).) If $u_1 w_2 \notin E$ and $u_2 w_1 \notin E$, then $G[u_1, u_2, w_1, w_2] \cong 2K_2$, a forbidden subgraph. If $u_1 w_2 \in E$ and $u_2 w_1 \in E$, then $G[u_1, u_2, w_1, w_2] \cong C_4$, also forbidden. Thus, we may assume $u_1 w_2 \in E$ and $u_2 w_1 \notin E$, as in Fig. 8.10(b).

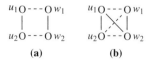

Figure 8.10

Because $\omega(G) \geq 3$, there exists a vertex $w \in W \setminus \{w_1, w_2\}$. If $u_1 w \notin E$, then either $u_2 w \notin E$, in which case $G[u_1, u_2, w_1, w] \cong 2K_2$, or $u_2 w \in E$, in which case $G[u_1, u_2, w, w_2] \cong C_4$. (Draw a picture.) Thus, $u_1 w \in E$ for all $w \in W \setminus \{w_1\}$. Let $W' = (W \setminus \{w_1\}) \cup \{u_1\}$ and $U' = V \setminus W' = (U \setminus \{u_1\}) \cup \{w_1\}$. Then W' is a clique and $o(W') = o(W) = \omega(G)$.

We claim that there exists a vertex $u \in U'$ such that $uw_1 \in E$ and $uu_1 \notin E$. Otherwise, because $u_2u_1 \in E$ but $u_2w_1 \notin E$, $G[U']$ would have fewer edges that $G[U]$, contradicting the choice of W. Moreover, $uu_2 \in E$ because otherwise $G[u, w_1, u_1, u_2] \cong 2K_2$, and $uw_2 \in E$ because otherwise $G[u, w_1, w_2, u_1, u_2] \cong C_5$. But that means $G[u, u_2, u_1, w_2] \cong C_4$. This final contradiction proves that $u_1u_2 \notin E$, in other words, that U is an independent set. ∎

8.15 Example. Let $\pi = (3, 2, 2, 2, 1)$. From the division of $F(\pi)$ exhibited in Fig. 8.11, we see that $\beta(\pi) \succeq_w \alpha(\pi)$, Thus π is graphic. Two nonisomorphic "realizations" of π are illustrated in the same figure. Note that G is chordal but H is not. Thus, whether a graph is chordal cannot be determined from its degree sequence. Because $\beta(\pi)$ contains too many boxes to majorize $\alpha(\pi)$, π is not a split partition. Hence, neither G nor H is a split graph. Indeed, $G[v_1, v_2, v_4, v_5] \cong 2K_2$ and H contains an induced subgraph isomorphic to C_4. Moreover, while G is chordal, $G^c \cong H$ is not. □

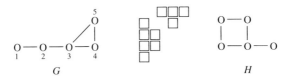

Figure 8.11

Interest in threshold and split graphs, and their relationship to chordal graphs, is all relatively recent.[4] The notion of a chordal graph, on the other hand, is much older.

Recall that the chromatic and clique numbers of G are related by the inequality $\chi(G) \geq \omega(G)$. This inequality is "sharp" in the sense that graphs exist for which equality holds. The "triangle," $G = C_3 = K_3$, is one example. On the other hand, for any positive integer N there exists a triangle-free graph G (i.e., $\omega(G) = 2$) with $\chi(G) \geq N$.[5] Of course, no graph with a cycle can be both triangle-free and chordal. Indeed, as we will soon see, $\chi(G) = \omega(G)$ whenever G is chordal.

8.16 Theorem[6]. *Let G be a graph on n vertices. If G does not contain an induced subgraph isomorphic to P_4, then $\chi(G) = \omega(G)$.*

---

[4] See, for example, S. Foldes and P. L. Hammer, Split graphs, *Proc. 8th S.E. Conference on Combinatorics, Graph Theory and Computing* (1977), 311–315; P. L. Hammer, T. Ibaraki, and B. Simeone, Threshold sequences, *SIAM J. Alg. Disc. Meth.* **2** (1981), 39–49; and P. L. Hammer and B. Simeone, The splittance of a graph, *Combinatorica* **1** (1981), 275–284.
[5] J. Mycielski, Sur le coloriage des graphs, *Colloq. Math.* **3** (1955), 161–162.
[6] Theorem 8.16 was first proved by D. Seinsche [On a property of the class of n-colorable graphs, *J. Combinatorial Theory* **16B** (1974), 191–193].

Proof. The proof is by induction on n. If $n = 1$, then $G = K_1$, and $\chi(G) = 1 = \omega(G)$. So, let G be a graph on $n \geq 2$ vertices that does not contain the forbidden subgraph P_4. By Theorem 1.19, one of G or G^c is disconnected. If G is not connected then, by the induction hypothesis, $\chi(H) = \omega(H)$ for every component H of G. Hence $\chi(G) = \max \chi(H) = \max \omega(H) = \omega(G)$, where both maxima are over the components H of G. If G is connected, then $G = H_1^c \vee H_2^c \vee \cdots \vee H_r^c$, where $H_1, H_2, \ldots, H_r$ are the components of G^c. By the induction hypothesis, $\chi(H_i^c) = \omega(H_i^c)$, $1 \leq i \leq r$. Thus, from Equations (6b) and (8b), $\chi(G) = \sum \chi(H_i^c) = \sum \omega(H_i^c) = \omega(G)$. ∎

8.17 Corollary. *Let G be a graph on n vertices. If G does not contain an induced subgraph isomorphic to P_4, then $\chi(H) = \omega(H)$ for every induced subgraph H of G.*

Proof. Let H be an induced subgraph of G. Because every induced subgraph of H is an induced subgraph of G, H does not contain the forbidden subgraph P_4. So, the result follows from Theorem 8.16. ∎

8.18 Definition. Graph G is *perfect* if $\chi(H) = \omega(H)$ for every induced subgraph H of G.[7]

While "P_4-free" graphs are perfect, not every perfect graph is P_4-free. Indeed, P_4 itself is a perfect graph. We now establish a pair of technical results necessary to the proof that every chordal graph is perfect. Recall that a subset $S \subset V(G)$ *separates* two nonadjacent vertices u and w of the connected graph G if u and w lie in different components of $G - S$. A *minimum separator* for u and w is a separating set S with $o(S) = \kappa_G(u, w)$.

8.19 Lemma. *Let G be a connected graph on n vertices. Then G is chordal if and only if each of its minimum separators is a clique.*

For the chordal graph G in Fig. 8.12, $\kappa_G(u, w) = 2$. The minimum separators of u and w are $\{v_1, v_2\}$, $\{v_2, v_3\}$, and $\{v_3, v_4\}$, each of which is a clique. While $\{v_1, v_4\}$ is not a clique, neither does it separate u and w.

Proof. Suppose G satisfies the minimum separator condition. Let $C = \langle u, v, w, x_1, \ldots, x_k \rangle$ be a cycle of G of length 4 or more. If $uw \in E(G)$, then C has a chord. If $uw \notin E(G)$, then any minimum separator for u and w must contain v and (at least) one of the x's, say x_t. Because any minimum separator is a clique, $vx_t \in E(G)$ is a chord of C.

---

[7] Perfect graphs were introduced by C. Berge [Perfect graphs, in *Six Papers in Graph Theory*, Indian Statistical Institute, Calcutta, 1963, pp 1–21], who conjectured that G is perfect if and only if G^c is perfect. Now known as the Perfect Graph Theorem, Berge's conjecture was proved by L. Lovász [A characterization of perfect graphs, *J. Combinatorial Theory* (B) **13** (1972), 95–98].

Figure 8.12

Conversely, suppose G is chordal. If $G = K_n$, then the minimum separator condition holds vacuously (i.e., it is true because it cannot be false). Otherwise, let u and w be a fixed but arbitrary pair of nonadjacent vertices of G. Let S be a minimum separator for u and w. If S is not a clique, it must contain two nonadjacent vertices, x and y. Let $G[U]$ and $G[W]$ be the components of $G - S$ containing u and w, respectively. (See Fig. 8.13.)

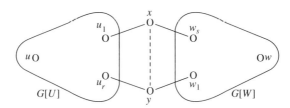

Figure 8.13

By the minimality of S, u and w are in the same component of $G - (S \setminus \{x\})$. Thus, $N_G(x) \cap U \neq \phi \neq N_G(x) \cup W$, and the same conclusion holds for the neighbors of y. It follows that the induced subgraphs $G[U \cup \{x, y\}]$ and $G[W \cup \{x, y\}]$ are connected. Let $P = [x, u_1, \ldots, u_r, y]$ be a shortest path in $G[U \cup \{x, y\}]$ from x to y and let $Q = [y, w_1, \ldots, w_s, x]$ be a shortest path in $G[W \cup \{x, y\}]$ from y to x. Because $C = <x, u_1, \ldots, u_r, y, w_1, \ldots, w_s>$ is a cycle of G of length not less than 4, it contains a chord. Because every path in G from a vertex in U to a vertex in W passes through S, none of the u's is adjacent to any of the w's. Together with the fact that P and Q are shortest paths, there is only one remaining possibility for a chord in C, namely an edge joining x to y, contradicting that $xy \notin E(G)$. ∎

8.20 Definition. Suppose $u \in V(G)$. If $d_G(u) = 0$ or $N_G(u)$ is a clique, then u is a *simplicial* vertex of G.

Suppose $G = (V, E)$ is a connected split graph on $n \geq 2$ vertices. Let $V = U \cup W$ be the disjoint union of a independent set U and a clique W. If $U \neq \phi$, then every element of U is a simplicial vertex. If $U = \phi$, then $G = K_n$ and all of its vertices are simplicial. On the other hand, if $G = C_n$ for some $n \geq 4$, then G has no simplicial vertices.

Chap. 8 Chordal Graphs

The proof that chordal graphs are perfect depends on the fact that they have simplicial vertices. Interestingly enough, the easiest way to prove that an arbitrary chordal graph has a simplicial vertex is to prove that it has two!

8.21 Lemma. *Let $G \neq K_n$ be a connected, chordal graph on $n \geq 3$ vertices. Then G has two nonadjacent simplicial vertices.*

Proof. The proof is by induction on n. If $n = 3$, then $G = P_3$, and the pendant vertices serve our purpose. So, assume $n \geq 4$ and let u and w be nonadjacent vertices of G. Let S be a minimum separator for u and w. By Lemma 8.19, S is a clique. Let $G[U]$ and $G[W]$ be the components of $G - S$ containing u and w, respectively. Define $H = G[U \cup S]$. Because every induced subgraph of H is an induced subgraph of G, H is chordal. If $H \cong K_r$, then $N_G(u)$ is a clique. In this case, let $x = u$. Otherwise, by induction, H contains two nonadjacent simplicial vertices, v_1 and v_2. Because $v_1 v_2 \notin E(G)$, v_1 and v_2 cannot both belong to S. Thus, (at least) one of them, say v_1, belongs to U. In this case, let $x = v_1$. In either case, $N_G(x) = N_H(x)$ is a clique. An analogous argument produces a vertex $y \in W$ such that $N_G(y)$ is a clique. Because $G[U]$ and $G[W]$ are different components of $G - S$, x and y are different and nonadjacent. ∎

8.22 Example. Figure. 8.14 shows three graphs in which the simplicial vertices are numbered. While the graphs in Figs. 8.14(a) and 8.14(c) are chordal, it is evident from Lemma 8.21 that the graph in Fig. 8.14(b) is not. (Find a chordless cycle.) □

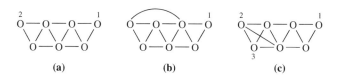

Figure 8.14

8.23 Theorem. *Every chordal graph is perfect.*

Proof. Because induced subgraphs of chordal graphs are chordal, it suffices to prove that $\chi(G) = \omega(G)$ for every chordal graph G. The proof is by induction on $n = o(V(G))$. If $n \leq 2$, then $G = K_1$, K_2, or $K_1 \oplus K_1$, each of which is chordal and perfect. If G is a disconnected chordal graph then, by induction, $\chi(C) = \omega(C)$ for each component of G in which case $\chi(G) = \omega(G)$ by Equations (6a) and (8a). If G is a connected chordal graph on $n \geq 3$ vertices, then either $G = K_n$, which is perfect, or, by Lemma 8.21, G contains a simplicial vertex v, in which case we consider $H = G - v$.

By the induction hypothesis, $\chi(H) = \omega(H)$, where $\omega(G) - 1 \leq \omega(H) \leq \omega(G)$. By the definition of simplicial vertex, $d_G(v) = o(N_G(v)) \leq \omega(H)$. If $d_G(v) <$

$\omega(H)$, then $\omega(G) = \omega(H)$. Moreover because the neighbors of v require at most $d_G(v) < \omega(H) = \chi(H)$ colors, any proper coloring of H can be extended to a proper coloring of G. Therefore, $\chi(G) = \chi(H) = \omega(H) = \omega(G)$.

If $d_G(v) = \omega(H)$, then $\{v\} \cup N_G(v)$ is a maximum clique of G, in which case $\omega(G) = \omega(H) + 1 = \chi(H) + 1 \geq \chi(G) \geq \omega(G)$. ∎

Suppose G is a connected, chordal graph on $n \geq 2$ vertices. It follows from Lemma 8.21 that G has a simplicial vertex, v_1. Because it is an induced subgraph of G, $G_1 = G - v_1$ is chordal, and because v_1 cannot be a cut vertex, G_1 is connected. If $G_1 \neq K_1$, then it has a simplicial vertex v_2, and $G_2 = G_1 - v_2 = G - v_1 - v_2$ is connected and chordal. Continuing in this manner, we come eventually to the point where $G_n \cong K_1$. Evidently, any chordal graph can be reduced to a collection of isolated vertices by successively deleting simplicial vertices.

8.24 Definition. Let G be a graph on n vertices. An ordering, $v_1, v_2, \ldots, v_n$, of the vertices of G is a *perfect elimination ordering* of G if v_i is a simplicial vertex of the induced subgraph $G[v_i, v_{i+1}, \ldots, v_n]$, $1 \leq i < n$.

8.25 Theorem. *Let G be a graph on n vertices. Then G is chordal if and only if it has a perfect elimination ordering.*

Proof. As we have just seen, every chordal graph has a perfect elimination ordering. The converse is proved by induction on n. Because every graph on $n \leq 3$ vertices is chordal, the proof is off to a good start. Let $v_1, v_2, \ldots, v_n$ be a perfect elimination ordering for G. Suppose $C = \langle w_1, w_2, \ldots, w_r \rangle$ is a cycle of G of length $r \geq 4$. If v_1 is a vertex of C, we may assume $v_1 = w_1$. Because v_1 is a simplicial vertex, its neighbors, w_2 and w_r are adjacent, that is, $w_2 w_r$ is a chord of C. Otherwise, C is a cycle in $G[v_2, v_3, \ldots, v_n]$, which, by induction, is a chordal graph. ∎

8.26 Corollary. *Suppose $v_1, v_2, \ldots, v_n$ is a perfect elimination ordering for the chordal graph G. Let d_i be the degree of vertex v_i, not in G but in the induced subgraph $G[v_i, v_{i+1}, \ldots, v_n]$, $1 \leq i < n$. Then the chromatic polynomial*

$$p(G, x) = x \prod_{i=1}^{n-1}(x - d_i).$$

It follows from Corollary 8.26 that every root (zero) of the chromatic polynomial of a chordal graph is a nonnegative integer.

Proof. If $n = 1$, then $p(G, x) = x$. If $n \geq 2$, let $u = v_1$ and $r = d_G(u) = d_1$. If $r = 0$ then u is an isolated vertex and $p(G, x) = xp(G - u, x) = (x -$

$d_1)p(G-u,x)$. If $r \geq 1$, then, since u is a simplicial vertex, $N_G(u)$ is a clique; that is, $G[N_G(u)] \cong K_r$. Therefore, G is an overlap of $G-u$ and K_{r+1} in K_r, and

$$p(G, x) = x^{(r+1)} p(G-u, x)/x^{(r)}$$
$$= (x - d_1)p(G-u, x).$$

Because $v_2, v_3, \ldots, v_n$ is a perfect elimination ordering for the chordal graph $G - u = G[v_2, v_3, \ldots, v_n]$, the result follows by induction. ∎

8.27 Example. Let G be either of the graphs from Fig. 8.7, reproduced (with a different vertex labeling) in Fig. 8.15. If v_i is the vertex numbered i, $1 \leq i \leq 7$, then $v_1, v_2, \ldots, v_7$ is a perfect elimination ordering for G, and $(d_1, d_2, \ldots, d_6) = (1, 1, 2, 2, 2, 1)$, where d_i is the degree of vertex v_i in $G[v_i, v_{i+1}, \ldots, v_n]$, $1 \leq i < n$. By Corollary 8.26, $p(G, x) = x(x-1)^3(x-2)^3$. In particular, G_1 and G_2 are nonisomorphic, chromatically equivalent graphs. □

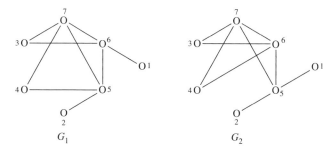

Figure 8.15

Example 8.27 can be generalized in the following way.

8.28 Corollary. *Suppose $\pi = (\pi_1, \pi_2, \ldots, \pi_n)$ is a split partition of $2m > 0$. Let G be a split graph with degree sequence $d(G) = \pi$. Then the chromatic polynomial*

$$p(G, x) = x^{(k)} \prod_{t=k+1}^{n} (x - \pi_t),$$

where $k = f(\pi)$ is the trace of π, and $x^{(k)}$ is the falling factorial function of degree k.

It follows from Corollary 8.28 that the chromatic polynomial of a split graph is completely determined by its degree sequence. In particular, two split graphs with the same degree sequence are chromatically equivalent.

Proof. Suppose $V(G) = \{v_1, v_2, \ldots, v_n\}$, where $d_G(v_i) = \pi_i$, $1 \leq i \leq n$. Since $k = f(\pi)$ we have, as in the proof of Theorem 8.9, that $W = \{v_i: 1 \leq i \leq k\}$ is a clique in G and $U = \{u_1, u_2, \ldots, u_{n-k}\}$ is an independent set, where $u_i = v_{k+i}$, $1 \leq i \leq n - k$. Because every vertex of U is a simplicial vertex of G, and every vertex of W is a simplicial vertex of $G - U$,

$$u_1, u_2, \ldots, u_{n-k}, v_1, v_2, \ldots, v_k$$

is a perfect elimination ordering for G. Because U is an independent set, the degree of u_i is π_{k+i}, both in G and in $G[u_i, u_{i+1}, \ldots, u_{n-k}, v_1, \ldots, v_k]$, $1 \leq i \leq n - k$. Because W is a clique, the degree of v_i in $G[v_i, v_{i+1}, \ldots, v_k]$ is $k - i$, $1 \leq i < k$. Thus, Corollary 8.28 is just the restriction of Corollary 8.26 to split graphs. ∎

Apart from applications within graph theory, chordal graphs have arisen in some unexpected places, for example, in studies of hyperplane arrangements[8] and in matrix completion problems.[9]

EXERCISES

1. Prove or disprove that β weakly majorizes $\alpha = (4, 3, 2, 1, 1)$ if $\beta =$
   (a) (5,5). (b) (4,4,4). (c) (6,6).
   (d) (3,3,3,2). (e) (5,2,1,1,1). (f) (6,5).

2. Exhibit the connected threshold graph corresponding to the shifted shape
   (a)  (b)

3. Find $\alpha(\pi)$ and $\beta(\pi)$ if $\pi =$
   (a) (5,4,3,3,2,1). (b) (5,4,2,2,2,2,1).
   (c) (4,2,2,2,2,2,1). (d) (5,4,2,2,1,1,1).

4. Exhibit a connected threshold graph whose degree sequence π Satisfies
   (a) $\alpha(\pi) = (6, 3, 1)$. (b) $\alpha(\pi) = (6, 4)$.

5. Which partitions in Exercise 3 are
   (a) graphic? (b) split? (c) threshold?

---

[8] P. H. Edelman and V. Reiner, Free hyperplane arrangements between A_{n-1} and B_n, *Math. Z.* **215** (1994), 347–365.
[9] R. Grone, C. R. Johnson, E. M. Sá, and H. Wolkowicz, Positive definite completions of partial hermitian matrices, *Linear Algebra Appl.* **58** (1984), 109–124.

Chap. 8 Chordal Graphs

6. As in the proof of Theorem 8.9, suppose K is a maximal clique in $G = (V, E)$. If $V \backslash K$ is an independant set, prove that K is a maximum clique. In other words, prove that $o(K) = \omega(G)$.

7. Find all threshold partitions τ that majorize the split partition $\pi = (4, 3, 3, 2, 1, 1)$.

8. Let $\pi \vdash 2m$ be a split partition
   (a) If $\tau \vdash 2m$ is a threshold partition. Show that $\tau \succ \pi$ if and only if $\beta(\pi) \succ \alpha(\tau) \succ \alpha(\pi)$.
   (b) If $\rho \vdash m$ satisfies $\beta(\pi) \succ \rho \succ \alpha(\pi)$, prove or disprove that $\rho = \alpha(\tau)$ for some threshold partition $\tau \vdash 2m$.

9. How many nonisomorphic, connected, threshold graphs have
   (a) 7 edges? (b) 8 edges? (c) 9 edges? (d) 10 edges?

10. Let G be a connected chordal graph. Suppose W is a maximal clique of G that does not contain a simplicial vertex of G. Prove that W is a vertex cut of G.

11. (Hammer and Simeone) Let $\pi \vdash 2m$ be a graphic partition of length $\ell(\pi) = n$. Let $k = k(\pi) = \max\{i : \pi_i \geq i - 1\}$. Show that π is a split partition if and only if

$$\sum_{i=1}^{k} \pi_i = k(k-1) + \sum_{i=k+1}^{n} \pi_i.$$

12. Find a perfect elimination ordering for the graph
    (a) in Fig. 8.14(a).
    (b) in Fig. 8.14(c).

13. Let G be the graph in Fig. 8.14(b). Find an induced subgraph of G that is connected and has more than one vertex, but which does not have a simplicial vertex.

14. Prove that G is chordal if and only if every induced subgraph of G has a simplicial vertex.

15. Let $G = (V, E)$ be a threshold graph, where V is the disjoint union of an independent set U and a clique W. Prove that there exists an ordering $u_1, u_2, \ldots, u_k$ of U such that $N_G(u_1) \supset N_G(u_2) \supset \cdots \supset N_G(u_k)$.

16. (Chvátal and Hammer) Suppose $G = (V, E)$ is a split graph whose vertex set has been partitioned as the disjoint union of an independent set U and a clique W. Suppose $N_G(u_1) \supset N_G(u_2) \supset \cdots \supset N_G(u_k)$, where $U = \{u_1, u_2, \ldots, u_k\}$. Prove that G is a threshold graph.

17. Prove that every bipartite graph is perfect.

18 Let G be the graph in Fig. 8.16.
   (a) Illustrate G^c, the complement of G.
   (b) Show that neither G nor G^c is chordal.

Figure 8.16

   (c) Show that neither G nor G^c is bipartite.
   (d) Show that, nevertheless, both G and G^c are perfect.

19 Let G be a graph on $n \geq 2$ vertices. Suppose G has a perfect elimination ordering. Let u be a fixed but arbitrary simplicial vertex of G and define $H = G - u$.
   (a) Prove that H has a simplicial vertex.
   (b) Prove that H has a perfect elimination ordering.

20 Prove that every interval graph (Chapter 7, Exercise 28) is chordal.

21 Suppose $G = (V, E)$ is a bipartite graph with color classes X and Y. Let $H = (V, F)$, where $F = E \cup X^{(2)}$, be the graph obtained from G by adding edges between every pair of vertices of X. Prove that H is a split graph.

22 (Read) Let G be the graph on 7 vertices obtained from K_6 by subdividing an edge.
   (a) Show that G has a chordless 4-cycles.
   (b) Show that $p(G, x) = x(x-1)(x-2)(x-3)^3(x-4)$.
   (c) Show that G need not be chordal for all of the roots of its chromatic polynomial to be integers.

23 Find a chordal graph H having the same chromatic polynomial as the graph G in Exercise 22.

24 (Sridharan and Balaji) Let G be a self–complementary graph on $n = 4k$ vertices. Prove that the following are equivalent:
   (i) G is chordal.
   (ii) G is split.
   (iii) $\omega(G) = 2k$.

25 Is there a difference between triangulated plane graphs and chordal plane graphs? Justify your answer.

Chap. 8 Chordal Graphs

26 Characterize those graphs G that are both bipartite and split.

27 Let $G = (V, E)$ be a split graph on $n \geq 2$ vertices. Show that V can be partitioned as the disjoint union of a *nonempty* independent set U and a *nonempty* clique W.

28 Suppose G is the graph in Fig. 8.14(a). Let v_i be the vertex labeled i, $1 \leq i \leq 2$.

   (a) Evaluate $\kappa_G(v_1, v_2)$.

   (b) Find a subset $S \subset V(G)$ such that

   (i) S separates v_1 and v_2, and

   (ii) S is not a clique.

   (c) Explain why part (b) does not disprove Lemma 8.19.

29 Prove or disprove that every tree is a split graph.

30 Suppose T is a tree on $n \geq 2$ vertices.

   (a) Explain why T is chordal.

   (b) Describe the simplicial vertices of T.

   (c) Let $T_1 = T$; if T_i has more than 2 vertices, let T_{i+1} be the tree obtained from T_i by deleting all s_i of its simplicial vertices. Show that the Balaban centric index, $B(T) = \sum s_i^2$.

   (d) Prove or disprove that $B(T)$ (not to do confused with $B(\pi)$) is equal to the number of perfect elimination orderings of T.

31 (Erdös–Gallai) Suppose $\pi \vdash 2m$, where $\ell(\pi) = n$. Prove that π is graphic if and only if

$$\sum_{i=1}^{k} \pi_i \leq k(k-1) + \sum_{i=k+1}^{n} \min\{k, \pi_i\}, \qquad 1 \leq k \leq n.$$

32 Suppose $\pi \vdash 2m$, where $\ell(\pi) = n$. Consider the $n \times (n-1)$ matrix $F'(\pi)$ whose (i, j)-entry is 1, if $j \leq \pi_i$, and 0, otherwise. If, for example, $\pi = (3, 2, 2, 2, 1)$, then

$$F'(\pi) = \begin{pmatrix} 1 & 1 & 1 & 0 \\ 1 & 1 & 0 & 0 \\ 1 & 1 & 0 & 0 \\ 1 & 1 & 0 & 0 \\ 1 & 0 & 0 & 0 \end{pmatrix}.$$

So, $F'(\pi)$ is just a variation of the Ferrers diagram, $F(\pi)$. As with $F(\pi)$, we divide $F'(\pi)$ into that portion, $A'(\pi)$, consisting of the entries on or above the main diagonal, and $B'(\pi)$, consisting of those entries below the main diagonal — that is,

$$B'(\pi) = \begin{matrix} & & 1 & 1 & 1 & 0 \\ & 1 & & 1 & 0 & 0 \\ 1 & 1 & & 0 & 0 \\ 1 & 1 & 0 & & 0 \\ 1 & 0 & 0 & 0 & \end{matrix} = A'(\pi).$$

Finally, let $M(\pi)$ be the $n \times n$ matrix with upper triangular part equal to $A'(\pi)$, lower triangular part equal to $B'(\pi)$, and main diagonal consisting entirely of zeros. If, for example, $\pi = (3, 2, 2, 2, 1)$, then

$$M(\pi) = \begin{pmatrix} 0 & 1 & 1 & 1 & 0 \\ 1 & 0 & 1 & 0 & 0 \\ 1 & 1 & 0 & 0 & 0 \\ 1 & 1 & 0 & 0 & 0 \\ 1 & 0 & 0 & 0 & 0 \end{pmatrix}.$$

(a) (Peled) Suppose $\pi \vdash 2m$ is a threshold partition. If G is the corresponding connected threshold graph, prove that $M(\pi) = A(G)$, the adjacency matrix of G.

(b) (Berge) Suppose $\pi \vdash 2m$, where $\ell(\pi) = n$. Denote by c_i the sum of the entries in column i of $M(\pi)$, $1 \leq i \leq n$. Prove that π is graphic if and only if

$$\sum_{i=1}^{k} \pi_i \leq \sum_{i=1}^{k} c_i, \quad 1 \leq k \leq n.$$

33 Prove that $\alpha(G)\omega(G) \geq n$ for every connected

   (a) threshold graph G on n vertices.

   (b) split graph G on n vertices.[10]

34 Recall (Chapter 2, Exercise 52) that G is critically k-chromatic if $k = \chi(G) > \chi(H)$ for every proper subgraph H of G. Let S be a vertex cut of the critically k-chromatic graph G. Prove that S is *not* a clique.[11] (*Hint*: Suppose S is a vertex cut of G. Let C_i, $1 \leq i \leq r$, be the components of $G - S$. Define $G_i = G[S \cup C_i]$. Consider colorings of the G_i that *agree* on S.)

---

[10] It was proved by L. Lovász [A characterization of perfect graphs, *J. Combinatorial Theory* (B) **13** (1972), 95–98] that G is perfect if and only if $\alpha(H)\omega(H) \geq o(V(H))$ for every induced subgraph H of G.

[11] Compare with Lemma 8.19.

Chap. 8 Chordal Graphs

35 Suppose G is a critically k-chromatic graph on $n \geq 3$ vertices. (See Exercise 34.)

  (a) Prove that G is 2-connected.

  (b) Suppose $\kappa(G) = 2$ and $S = \{u, v\}$ is a vertex cut. Prove that $uv \notin E(G)$.

36 Give a proof of the formula $p(K_n, x) = x^{(n)}$ based on Corollary 8.26.

37 Illustrate Corollary 8.26 by using it to obtain the chromatic polynomial for the graph

  (a) in Fig. 8.14(a).

  (b) in Fig. 8.14(c).

38 Explain why Corollary 8.26 cannot be used to obtain the chromatic polynomial for the graph in Fig. 8.14(b).

39 Let A_n be the unique, connected antiregular graph on $n \leq 2$ vertices. (See Chapter 2, Exercise 36.) Prove that

$$p(A_n, x) = \begin{cases} x[(x-1)(x-2)\ldots(x-s)]^2, & n \text{ odd} \\ x[(x-1)(x-2)\ldots(x-s+1)]^2(x-s), & n \text{ even,} \end{cases}$$

where $s = \lfloor n/2 \rfloor$, the integer part of $n/2$.

40 Exhibit two nonisomorphic connected graphs with degree sequence $(5, 4, 2, 2, 2, 2, 1)$.

9

Oriented Graphs

Recall that graph $G = (V, E)$ consists of a nonempty set V of n elements, together with a (possibly empty) subset E of $V^{(2)}$, the family of all $C(n, 2)$ 2-element subsets of V. Not to be confused with the set $V \times V = \{(v_1, v_2) : v_1, v_2 \in V\}$ of all n^2 ordered pairs of vertices, the elements of $V^{(2)}$ are *unordered*.

9.1 Definition. An *orientation* of $G = (V, E)$ is a function $f : E \to V \times V$ such that, for all $e = \{u, w\} \in E$, $f(e)$ is one of (u, w) or (w, u).

By the Fundamental Counting Principle, a graph with $m \geq 1$ edges has 2^m orientations. Laying common sense aside, we adopt the convention that the number of orientations of K_n^c is one.

An *oriented graph* is a graph with a nonempty set of edges and a prescribed orientation. Suppose $G = (V, E)$ is a graph oriented by f. Let $e = \{x, y\} \in E$. If $f(e) = (x, y)$, then vertex x is the *tail* of the *oriented edge* $e = (x, y)$, and vertex y is its *head*. Consistent with this language, $e = (x, y)$ is typically illustrated by a *directed* arc, or *arrow,* from x to y.[1]

9.2 Example. Suppose a city softball league organizes a championship tournament in which each of five qualifying teams plays one game with each of the other qualifiers. Perhaps the teams are *seeded* (ranked) 1–5. The results of such a tournament can be illustrated by an orientation of K_5 in which oriented edge $e = (u, w)$ indicates that team u defeated team w. (See Fig. 9.1.) Indeed, this has become such a widely used model for "round-robin tournaments" that oriented complete graphs have come, themselves, to be known as *tournaments*. □

Figure 9.1

---

[1] An oriented graph is a special kind of *directed* graph in which, for all $\{u, v\} \in V^{(2)}$, at most one of (u, v) and (v, u) can be an edge.

171

9.3 Definition. Let G be an oriented graph. A *directed uv-path* in G is a path $[u = v_0, v_1, \ldots, v_r = w]$ where (v_{i-1}, v_i) is an oriented edge of G, $1 \leq i \leq r$. A *directed* cycle in G is a cycle $\langle v_1, v_2, \ldots, v_r \rangle$ in which (v_i, v_{i+1}), $1 \leq i < r$, and (v_r, v_1) are oriented edges of G.

9.4 Example. The oriented graph illustrated in Fig. 9.1 contains many directed cycles — for example, $\langle 3, 5, 4 \rangle$, $\langle 2, 4, 3 \rangle$, $\langle 1, 3, 2 \rangle$, $\langle 1, 4, 3, 2 \rangle$, and even a spanning directed cycle, $\langle 1, 4, 3, 5, 2 \rangle$. The oriented graph in Fig. 9.2 has no directed cycles. □

Figure 9.2

9.5 Definition. An *acyclic* orientation of G is one in which there are no directed cycles.

Because it has no cycles at all, every orientation of a tree is acyclic. What about K_5? Of its $2^{10} = 1024$ orientations, how many are acyclic? The answer is an easy consequence of a remarkable result due to R. P. Stanley.[2]

9.6 Stanley's Theorem. *Let G be a graph on n vertices. Then the number of acyclic orientations of G is $(-1)^n p(G, -1)$, where $p(G, x)$ is the chromatic polynomial of G.*

Proof. Let $\rho(G) = (-1)^n p(G, -1)$. If G has no edges, then $p(G, x) = x^n$ and $\rho(G) = (-1)^n (-1)^n = 1$. If $e = \{u, v\} \in E(G)$, then, because $-(-1)^{n-1} = (-1)^n$, chromatic reduction yields

$$\rho(G) = (-1)^n [p(G - e, -1) - p(G/e, -1)]$$
$$= \rho(G - e) + \rho(G/e). \tag{58}$$

If $e = \{u, v\}$, then recall that $G|e$ is the multigraph obtained from $G - e$ by identifying vertices u and v. (See, e.g., Fig. 9.3.) Because multiple adjacencies are irrelevant to proper colorings, $p(G|e, x) = p(G/e, x)$. (That is why the contraction, G/e, was introduced in the first place.) However, because their orientations may differ, multiple edges are *not* irrelevant here. For our present purposes, it will be useful to take a step backward and write Equation (58) in the form

$$\rho(G) = \rho(G - e) + \rho(G|e).$$

---

[2] R. P. Stanley, Acyclic orientations of graphs, *Discrete Math.* **5** (1973), 171–178.

Chap. 9 Oriented Graphs

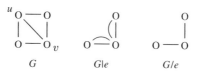

Figure 9.3

Let $c(G)$ to be the number of acyclic orientations of G. If G has no edges then, by convention, $c(G) = 1$. Thus, to prove that $c(G)$ and $p(G)$ are identical, it suffices to show that they satisfy the same recurrence relation, to show that $c(G) = c(G - e) + c(G|e)$, $e \in E(G)$.

Let $e = \{u, v\}$ be a fixed but arbitrary (unoriented) edge of G. If f is an orientation of G, denote its restriction to $G - e$ by f_e. If f is acyclic, then f_e is acyclic. That part is straightforward. The reverse is complicated by the fact that an orientation of $G - e$ can be extended to G (strictly speaking, to $E(G)$) in two different ways.

Denote by $\overline{f}$ the orientation of G obtained from f by switching (only) the orientation of our fixed but arbitrary edge e. If, for example, $f(e) = (u, v)$, then $\overline{f}(e) = (v, u)$. Apart from e, f and $\overline{f}$ agree on every edge of G. In particular, $f_e = \overline{f}_e$.

Let g be a fixed but arbitrary acyclic orientation of $G - e$. Let f be the (unique) extension of g to G that satisfies $f(e) = (u, v)$. Then the f-orientation of G is acyclic if and only if the g-orientation of $G - e$ does not afford a directed vu-path. Similarly, the $\overline{f}$-orientation of G is acyclic if and only if g does not contain a directed uv-path. Because g is acyclic, it cannot afford both a vu-path and a uv-path. Thus, every acyclic orientation of $G - e$ extends either to one or to two acyclic orientations of G.

Let $C(G)$ be the set of acyclic orientations of G. (So, $c(G) = \mathrm{o}(C(G))$.) If $B(G) = \{f \in C(G): \overline{f}$ is (also) acyclic$\}$, then $C(G) \setminus B(G) = \{f \in C(G): \overline{f}$ is not acyclic$\}$. Our observations up to this point can be summarized by the formula

$$c(G - e) = \mathrm{o}(B(G))/2 + \mathrm{o}(C(G) \setminus B(G))$$
$$= c(G) - \mathrm{o}(B(G))/2,$$

because $\mathrm{o}(C(G) \setminus B(G)) = \mathrm{o}(C(G)) - \mathrm{o}(B(G))$.

What about $G|e$? This case is easier. The restriction of f to $G|e$ is acyclic if and only if both f and $\overline{f}$ are acyclic orientations of G. Thus, $c(G|e) = \mathrm{o}(B(G))/2$. Putting the two formulas together, we obtain

$$c(G - e) + c(G|e) = c(G). \qquad \blacksquare$$

9.7 Example. Recall that $p(K_n, x) = x^{(n)}$, the falling factorial function. Thus, $p(K_5, x) = x(x - 1)(x - 2)(x - 3)(x - 4)$. Of the 1024 orientations of K_5, $(-1)^5 p(K_5, -1) = 5! = 120$ are acyclic. If T is a tree on n vertices, then T has $m =$

$n-1$ edges and 2^{n-1} orientations. From Theorem 2.30, $p(T, x) = x(x-1)^{n-1}$. Hence, T has $(-1)^n p(T, -1) = 2^{n-1}$ acyclic orientations, confirming our earlier observation that every orientation of a tree is acyclic.

The cycle, C_n, has $m = n$ edges and 2^n orientations. Because only the "clockwise" and "counterclockwise" orientations afford directed cycles, C_n must have $2^n - 2$ acyclic orientations. From Chapter 2, Exercise 26, $p(C_n, x) = (x-1)^n + (-1)^n(x-1)$ and, sure enough, $(-1)^n p(C_n, -1) = 2^n - 2$. □

9.8 Definition. Suppose $G = (V, E)$ is an oriented graph with vertex set $V = \{v_1, v_2, \ldots, v_n\}$ and edge set $E = \{e_1, e_2, \ldots, e_m\}$. Let $Q(G) = (q_{ij})$ be the $n \times m$ matrix defined by $q_{ij} = 1$ if v_i is the head of e_j, -1 if v_i is the tail of e_j, and 0 otherwise. Then $Q(G)$ is an *oriented, vertex-edge incidence matrix* for G.

9.9 Example. Note that $Q(G)$ depends not only on G but on the orientation of G and on the numberings of its vertices and edges. If G_1 and G_2 are the oriented graphs illustrated in Fig. 9.4, then

$$Q(G_1) = \begin{pmatrix} 1 & 0 & 0 \\ 0 & -1 & 0 \\ 0 & 1 & 1 \\ -1 & 0 & -1 \end{pmatrix} \quad \text{and} \quad Q(G_2) = \begin{pmatrix} -1 & 1 & 0 & 0 & 0 \\ 1 & 0 & -1 & -1 & 0 \\ 0 & 0 & 1 & 0 & 1 \\ 0 & -1 & 0 & 1 & -1 \end{pmatrix}.$$

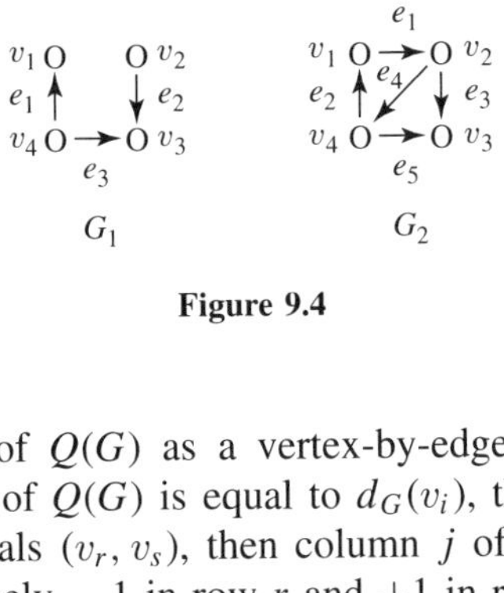

Figure 9.4

It is useful to think of $Q(G)$ as a vertex-by-edge matrix. The number of nonzero entries in row i of $Q(G)$ is equal to $d_G(v_i)$, the degree of vertex v_i. If the oriented edge e_j equals (v_r, v_s), then column j of $Q(G)$ contains precisely two nonzero entries, namely, -1 in row r and $+1$ in row s. □

Denote by Q^t the *transpose* of $Q = Q(G)$. Then Q^t is the $m \times n$ (edge-by-vertex) matrix whose (i,j)-entry is q_{ji}, the (j,i)-entry of Q. As it turns out, we are less interested in Q than in the (vertex-by-vertex) matrix QQ^t.

9.10 Example. If $Q = Q(G_1)$ from Example 9.9, then

$$QQ^t = \begin{pmatrix} 1 & 0 & 0 & -1 \\ 0 & 1 & -1 & 0 \\ 0 & -1 & 2 & -1 \\ -1 & 0 & -1 & 2 \end{pmatrix}.$$

If $Q = Q(G_2)$ from Example 9.9, then

$$QQ^t = \begin{pmatrix} 2 & -1 & 0 & -1 \\ -1 & 3 & -1 & -1 \\ 0 & -1 & 2 & -1 \\ -1 & -1 & -1 & 3 \end{pmatrix}.$$

Pause for a minute to confirm that these matrix products have been computed correctly and, at the same time, guess at the relationship between G and the entries of QQ^t. □

9.11 Definition. Let G be a graph with vertex set $V(G) = \{v_1, v_2, \ldots, v_n\}$. The *Laplacian matrix* of G is

$$L(G) = D(G) - A(G),$$

where $D(G) = \text{diag}(d_G(v_1), d_G(v_2), \ldots, d_G(v_n))$ is the diagonal matrix of vertex degrees, and $A(G)$ is the adjacency matrix.

9.12 Example. If G is the graph in Fig. 9.5, then

$$L(G) = \begin{pmatrix} 4 & -1 & -1 & -1 & -1 \\ -1 & 2 & -1 & 0 & 0 \\ -1 & -1 & 3 & -1 & 0 \\ -1 & 0 & -1 & 2 & 0 \\ -1 & 0 & 0 & 0 & 1 \end{pmatrix}.$$ □

Figure 9.5

9.13 Theorem. *Let G be a graph with vertex set $V(G) = \{v_1, v_2, \ldots, v_n\}$. Let $Q = Q(G)$ be an oriented vertex-edge incidence matrix for G. Then, independently of the orientation of G and of the numbering of its edges, $L(G) = QQ^t$.*

Proof. From the definitions of transpose and matrix multiplication, the (i,j)-entry of QQ^t is

$$\sum_{r=1}^{m}(Q)_{ir}(Q^t)_{rj} = \sum_{r=1}^{m} q_{ir}q_{jr}. \tag{59}$$

If $i = j$, then $q_{ir}q_{jr} = q_{ir}^2$, and (59) is the sum of the squares of the entries in row i of $Q(G)$. Since $q_{ir} = \pm 1$ when v_i is incident with e_r, and 0 otherwise, the sum over r of q_{ir}^2 is precisely $d_G(v_i)$. If $i \neq j$, then $q_{ir}q_{jr} \neq 0$ if and only if $q_{ir} \neq 0 \neq q_{jr}$, if and only if $\{v_i, v_j\} = e_r \in E(G)$, if and only if $q_{ir}q_{jr} = -1$. Hence, provided $i \neq j$, the (i,j)-entry of QQ^t is -1 when $\{v_i, v_j\} \in E(G)$, and 0 otherwise. ∎

9.14 Example. Let G be a graph with vertex set $V(G) = \{v_1, v_2, \ldots, v_n\}$, and Laplacian matrix $L(G) = D(G) - A(G)$. Because the number of 1's in row i of $A(G)$ is $d_G(v_i)$, the sum of the entries in row i of $L(G)$ is zero, $1 \leq i \leq n$. This is equivalent to

$$L(G)Y_n = 0,$$

where Y_n is the $n \times 1$ matrix, each of whose entries is 1. The following useful facts flow from this equation: Writing it in the form $L(G)Y_n = \lambda Y_n$ shows that Y_n is an eigenvector of $L(G)$ corresponding to the eigenvalue $\lambda = 0$. In particular, $L(G)$ is a singular matrix; that is, det $(L(G)) = 0$. □

Suppose X is an eigenvector of $L(G)$ corresponding to an arbitrary eigenvalue λ. Then X is a nonzero, $n \times 1$ matrix and $L(G)X = \lambda X$. Therefore, $X^t L(G) X = X^t(L(G)X) = X^t(\lambda X) = \lambda(X^t X) = \lambda \|X\|^2$, where $\|X\|^2$ is the "dot product" of X with itself — that is, the sum of the squares of the entries of X. From Theorem 9.13, on the other hand, $X^t L(G) X = X^t(QQ^t)X = (X^t Q)(Q^t X) = (Q^t X)^t(Q^t X) = \|Q^t X\|^2$. Combining these identities, we find that

$$\lambda = \|Q^t X\|^2 / \|X\|^2 \geq 0.$$

Evidently, the eigenvalues of $L(G)$ are all nonnegative.[3]

9.15 Definition. Suppose G is a graph on n vertices. Denote the *spectrum* of $L(G)$ by $s(G) = (\lambda_1(G), \lambda_2(G), \ldots, \lambda_n(G))$, where the eigenvalues of $L(G)$ are ordered so that

$$\lambda_1(G) \geq \lambda_2(G) \geq \cdots \geq \lambda_n(G) = 0. \tag{60}$$

While it does not depend on the numbering of the edges (or on any orientation), $L(G)$ does depend on the vertex numbering of G. If the vertex numbers are permutted, the effect on $L(G)$ is a corresponding permutation of its rows and columns. More generally, as with the adjacency matrix, graphs G_1 and G_2 are isomorphic if and only if there is a permutation matrix P (corresponding to the isomorphism) such that

$$L(G_1) = P^{-1} L(G_2) P.$$

---

[3] Every characteristic root of a real symmetric matrix is known to be real. If they are all nonnegative, then the symmetric matrix is said to be *positive semidefinite*.

Because matrix similarity preserves eigenvalues, it follows from this identity that the Laplacian spectrum is a graph invariant; that is, $G_1 \cong G_2$ only if $s(G_1) = s(G_2)$.[4]

9.16 Example. Let G be the graph in Fig. 9.5 whose Laplacian matrix was given in Example 9.12. Computations show that the characteristic polynomial of $L(G)$ is

$$\det(xI - L(G)) = x^5 - 12x^2 + 49x^3 - 78x^2 + 40x$$
$$= x(x-1)(x-2)(x-4)(x-5), \qquad (61)$$

so $s(G) = (5, 4, 2, 1, 0)$. □

Relations between eigenvalues and other graph invariants comprise an important part of what has come to be known as *algebraic graph theory*. This is a story that combines early results from matrix theory and electrical circuit theory.

Let A be a real $n \times n$ matrix. Denote by A_{ij} the $(n-1) \times (n-1)$ submatrix of A obtained by deleting its ith row and jth column. The *adjugate* (or *classical adjoint*) of A is the $n \times n$ matrix $A^\dagger$ whose (i,j)-entry is $(-1)^{i+j} \det(A_{ji})$. The result that makes adjugates worth knowing about is

$$AA^\dagger = \det(A) I_n. \qquad (62)$$

It follows from Equation (62) that $A^{-1} = (1/\det(A))A^\dagger$, whenever $\det(A) \neq 0$, a formula that is not of much use when A is the singular matrix $L(G)$. Substituting $A = L(G)$ into Equation (62), however, produces $L(G)L(G)^\dagger = 0$, an identity that can be interpreted as $L(G)C = 0$ for every column C of $L(G)^\dagger$.

Recall that the rank of a matrix is equal to the size of a largest square submatrix having a nonzero determinant. Because $\det(L(G)) = 0$, rank $L(G) < n$. From the definition of $L(G)^\dagger$, its n^2 entries are precisely the determinants of all possible $(n-1) \times (n-1)$ submatrices of $L(G)$. It follows that rank $L(G) \leq n-2$, if and only if $L(G)^\dagger = 0$, if and only if $C = 0$ for every column of $L(G)^\dagger$. If, on the other hand, the rank of $L(G)$ is $n-1$, then the dimension of its null space is 1, meaning that $L(G)X = 0$ if and only if X is a multiple of the eigenvector Y_n, if and only if the entries of X are all the same. Evidently, regardless of the rank of $L(G)$,

$$L(G)^\dagger = \begin{pmatrix} a & b & c & \ldots & d \\ a & b & c & \ldots & d \\ & & \ldots & & \\ a & b & c & \ldots & d \end{pmatrix}, \qquad (63)$$

---

[4] An $n \times n$ matrix of integers, M, is *unimodular* if it is invertible and M^{-1} is an integer matrix. (It is not hard to see that M is unimodular if and only if $\det(M) = \pm 1$.) Integer matrices A and B are *congruent* if there is a unimodular matrix M such that $A = M^t B M$. Since $P^{-1} = P^t$, $G_1 \cong G_2 \iff L(G_1) = P^t L(G_2) P \implies L(G_1)$ and $L(G_2)$ are congruent. A partial converse was obtained by W. Watkins [Unimodular congruence of the Laplacian matrix of graphs, *Linear Algebra Appl.* **201** (1994), 43–49]: If $L(G_1)$ and $L(G_2)$ are congruent and if G_1 is 3-connected, then $G_1 \cong G_2$.

where $a, b, c, \ldots$, and d are constants. Because the adjugate of a symmetric matrix is symmetric, the numbers in the first row of $L(G)^\dagger$ are equal to the corresponding numbers in its first column. In other words, Equation (63) implies that every entry of $L(G)^\dagger$ is the same!

9.17 Theorem. *Let G be a graph on n vertices. Then there exists an integer, $t(G)$, depending only on G, such that*

$$t(G) = (-1)^{i+j} \det(L(G)_{ij}), \quad 1 \le i, j \le n. \tag{64}$$

Moreover, $t(G) = 0$ if and only if rank $L(G) < n - 1$.

Proof. We have already proved the existence of a number $t(G)$ that satisfies Equation (64) and the condition that $t(G) = 0$ if and only if rank $L(G) < n - 1$. It remains to observe that, because $L(G)_{ij}$ is an integer matrix, its determinant is an integer. ∎

When an integer emerges from a combinatorial setting, it is natural to expect that it is counting something. Recall that a spanning tree of G is a spanning subgraph that is a tree.

9.18 Example. The graph in Fig. 9.5 has a total of eight spanning trees; they are illustrated in Fig. 9.6. □

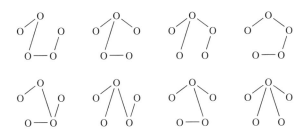

Figure 9.6

With a little imagination, the next result can be found in an 1847 article by G. Kirchhoff.[5]

9.19 Matrix-Tree Theorem. *Let G be a graph on n vertices. Then $t(G)$ is the number of spanning trees in G.*

Proof Sketch. By Theorem 9.17, it suffices to show that $\det(L(G)_{nn})$ is equal to the number of spanning trees of G. Applying the Cauchy–Binet Theorem, one

---

[5] G. Kirchhoff, Über die Auflösung der Gleichungen, auf welche man bei der Untersuchung der linearen Verteilung galvanischer Ströme geführt wird, *Ann. Phys. Chem.* **72** (1847), 497–508.

Chap. 9   Oriented Graphs

of the classics of nineteenth-century matrix theory, to the factorization $L(G) = QQ^t$, one finds that $\det(L(G)_{nn})$ is the sum of the squares of the determinants of the $(n-1) \times (n-1)$ submatrices of Q lying in its first $n-1$ rows and all possible selections of $n-1$ of its columns. Our theorem now follows from the fact that any such subdeterminant of Q is equal to 1, -1, or 0, with its value being nonzero if and only if the corresponding $n-1$ columns of Q correspond to the edges in a spanning tree of G. (See Exercise 34, below.) ∎

9.20 Example.   Consider, once again, the graph G illustrated in Fig. 9.5. Note that Theorem 9.19 concerns the number of *different* spanning trees of G. The fact that there are numerous isomorphisms among the eight trees in Fig. 9.6 is irrelevant to the computation of $t(G)$. From Example 9.12, we have

$$L(G) = \begin{pmatrix} 4 & -1 & -1 & -1 & -1 \\ -1 & 2 & -1 & 0 & 0 \\ -1 & -1 & 3 & -1 & 0 \\ -1 & 0 & -1 & 2 & 0 \\ -1 & 0 & 0 & 0 & 1 \end{pmatrix}.$$

Hence, for example,

$$L(G)_{11} = \begin{pmatrix} 2 & -1 & 0 & 0 \\ -1 & 3 & -1 & 0 \\ 0 & -1 & 2 & 0 \\ 0 & 0 & 0 & 1 \end{pmatrix}.$$

Expanding first along its last row, the determinant of $L(G)_{11}$ is

$$\det \begin{pmatrix} 2 & -1 & 0 \\ -1 & 3 & -1 \\ 0 & -1 & 2 \end{pmatrix} = 2 \det \begin{pmatrix} 3 & -1 \\ -1 & 2 \end{pmatrix} + \det \begin{pmatrix} -1 & -1 \\ 0 & 2 \end{pmatrix}$$

$$= 2(6-1) - 2 = 8.$$

Together with Example 9.18, this confirms the interpretation of $t(G)$ given in Theorem 9.19. To confirm Theorem 9.17, let's calculate, say, the $(1, 3)$-entry of $L(G)^\dagger$. Deleting row 3 and column 1 from $L(G)$ and expanding determinants along the last row and then again along the last row, we obtain

$$t(G) = (-1)^4 \det \begin{pmatrix} -1 & -1 & -1 & -1 \\ 2 & -1 & 0 & 0 \\ 0 & -1 & 2 & 0 \\ 0 & 0 & 0 & 1 \end{pmatrix}.$$

$$= \det \begin{pmatrix} -1 & -1 & -1 \\ 2 & -1 & 0 \\ 0 & -1 & 2 \end{pmatrix}$$

$$= -(-1)\det\begin{pmatrix} -1 & -1 \\ 2 & 0 \end{pmatrix} + 2\det\begin{pmatrix} -1 & -1 \\ 2 & -1 \end{pmatrix}$$
$$= 2 + 2(3) = 8. \qquad \square$$

Recall that the coefficient of x in the characteristic polynomial of an $n \times n$ matrix A is

$$(-1)^{n-1} \sum_{i=1}^{n} \det(A_{ii}).$$

In the characteristic polynomial of $L(G)$, the coefficient of x is $(-1)^{n-1} \times n \times t(G)$. For the graph in Fig. 9.5, we have $(-1)^4 \times 5 \times 8 = 40$, consistent with Equation (61).

On the other hand, the coefficient of x in the product

$$(x - \lambda_1)(x - \lambda_2) \cdots (x - \lambda_n)$$

is $(-1)^{n-1}$ times the sum of the products of the λ's taken $n-1$ at a time. If $A = L(G)$, then, because $\lambda_n = \lambda_n(G) = 0$, only one of these products survives, namely, $\lambda_1(G) \times \lambda_2(G) \times \cdots \times \lambda_{n-1}(G)$. (Confirm that this formula is consistent with Equation (61).) Equating these two ways of computing the coefficient of x in $\det(xI_n - L(G))$ yields

$$nt(G) = \prod_{i=1}^{n-1} \lambda_i(G). \qquad (65)$$

9.21 Corollary. *Let G be a graph on n vertices. Then $\lambda_{n-1}(G) > 0$ if and only if G is connected.*

Proof. The result follows from (60), (65), and the fact that G is connected if and only if $t(G) \neq 0$. ∎

Corollary 9.21 suggests that $\lambda_{n-1}(G)$ might be viewed as another quantitative measure of connectivity.

9.22 Definition. The *algebraic connectivity* of a graph G on n vertices is $a(G) = \lambda_{n-1}(G)$.

The algebraic connectivity was introduced by M. Fiedler,[6] who showed, among a great many other things, that

$$a(G) \leq \kappa(G), \qquad (66)$$

provided $G \neq K_n$.

---

[6] M. Fiedler, Algebraic connectivity of graphs, *Czech Math. J.* **23** (1973), 298–305. Eigen*vectors* affording $a(G)$ have proved useful in the design of algorithms for the efficient use of parallel processors. See, for example, S. T. Barnard and H. D. Simon, A fast multilevel implementation of recursive spectral bisection for partitioning unstructured problems, *Concurrency: Practice and Experience* **6** (1994), 101–117 and B. Hendrickson and R. Leland, An improved spectral graph partitioning algorithm for mapping parallel computations, *SIAM J. Scientific Computing* **16** (1995), 452–469.

Suppose $G = H_1 \oplus H_2$. Then, with a suitable numbering of its vertices, $L(G) = L(H_1) \oplus L(H_2)$, the direct sum of the Laplacian matrices corresponding to H_1 and H_2, that is,

$$L(H_1 \oplus H_2) = L(H_1) \oplus L(H_2). \tag{67}$$

In particular, $s(G)$ is obtained by rearranging, into nonincreasing order, the numbers from $s(H_1)$ and $s(H_2)$. Putting this discussion together with Corollary 9.21 yields the following.

9.23 Corollary. *Let G be a graph on n vertices. Then the multiplicity of 0 as an eigenvalue of $L(G)$ is equal to $\xi(G)$, the number of connected components of G.*

Suppose X is an eigenvector of $L(G)$ corresponding to the eigenvalue μ. Because $L(G)$ is symmetric, $L(G)X = \mu X$ implies $\mu X^t = X^t L(G)$. If Y is an eigenvector afforded by ν, then

$$\mu X^t Y = (X^t L(G))Y$$
$$= X^t (L(G)Y)$$
$$= \nu X^t Y,$$

from which we obtain $(\mu - \nu)X^t Y = 0$. If $\mu \neq \nu$, it must be that the dot product, $Y \cdot X = X^t Y = 0$: Eigenvectors of $L(G)$, corresponding to distinct eigenvalues, are orthogonal. More generally, given $s(G) = (\lambda_1, \lambda_2, \ldots, \lambda_n)$, there exist eigenvectors $X_1, X_2, \ldots, X_n$ such that $L(G)X_i = \lambda_i X_i$, $1 \leq i \leq n$, and $X_i \cdot X_j = 0$ for all $i \neq j$.[7]

Consider $G^c = (V, V^{(2)} \setminus E)$, the complement of $G = (V, E)$. If we use the same vertex numbering for both graphs, then

$$L(G) + L(G^c) = nI_n - J_n, \tag{68}$$

where J_n is the $n \times n$ matrix each of whose entries is 1. Suppose $\mu = \lambda_i(G)$ for some $i < n$. By the previous observations, there exists a (nonzero) $n \times 1$ matrix X such that $L(G)X = \mu X$ and $(Y_n)^t X = X \cdot Y_n = 0$. Because $(Y_n)^t$ is the ith row of J_n, $1 \leq i \leq n$, it must be that $J_n X = 0$. Together with Equation (68), this gives $L(G)X + L(G^c)X = nX$. Because $L(G)X = \mu X$, this last equation is equivalent to $L(G^c)X = (n - \mu)X$. It seems we have proved the following.

9.24 Theorem. *If G is a graph on n vertices, then $\lambda_{n-i}(G^c) = n - \lambda_i(G)$, $1 \leq i < n$; that is, if $s(G) = (\lambda_1, \lambda_2, \ldots, \lambda_{n-1}, 0)$, then $s(G^c) = (n - \lambda_{n-1}, \ldots, n - \lambda_2, n - \lambda_1, 0)$.*

9.25 Corollary. *Let G and H be graphs on disjoint sets of t and r vertices, respectively. If $s(G) = (\nu_1, \ldots, \nu_{t-1}, 0)$ and $s(H) = (\mu_1, \ldots, \mu_{r-1}, 0)$, then $s(G \vee$*

---
[7] This is a version of the *Spectral Theorem* for symmetric matrices.

H) consists of the numbers $r + t$,

$$t + \mu_1, \ldots, t + \mu_{r-1},$$
$$r + \nu_1, \ldots, r + \nu_{t-1},$$

and 0, arranged in nonincreasing order.

Proof. From Theorem 9.24, $s(G^c) = (t - \nu_{t-1}, \ldots, t - \nu_1, 0)$ and $s(H^c) = (r - \mu_{r-1}, \ldots, r - \mu_1, 0)$. From Equation (67), $s(G^c \oplus H^c)$ is some rearrangement of

$$0, r - \mu_1, \ldots, r - \mu_{r-1},$$
$$t - \nu_1, \ldots, t - \nu_{t-1},$$

and 0. Together with $G \vee H = (G^c \oplus H^c)^c$ and the fact that $n = r + t$, the result follows from another application of Theorem 9.24. ∎

9.26 Example. Consider the graphs G_1, G_2, H_1, and H_2 shown in Fig. 9.7. Observe that a pair of numbered vertices is adjacent in G_1 if and only if

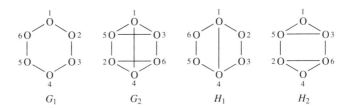

Figure 9.7

the same numbered pair is *not* adjacent in G_2. This proves, by the numbers, that $G_2 \cong G_1^c$. Similarly, $H_2 \cong H_1^c$. It can be shown (see Exercise 31, below) that $s(G_1) = (4, 3, 3, 1, 1, 0)$ and $s(H_1) = (5, 3, 3, 2, 1, 0)$. By Theorem 9.24, $s(G_2) = (5, 5, 3, 3, 2, 0)$ and $s(H_2) = (5, 4, 3, 3, 1, 0)$. From Equation (67),

$$S(G_1 \oplus G_2) = (5, 5, 4, 3, 3, 3, 3, 2, 1, 1, 0, 0)$$
$$= S(H_1 \oplus H_2).$$

Thus, while they are not isomorphic, $G = G_1 \oplus G_2$ and $H = H_1 \oplus H_2$ have the same Laplacian spectrum. By Corollary 9.25, $G^c = (G_1 \oplus G_2)^c = G_1^c \vee G_2^c = G_2 \vee G_1 = G_1 \vee G_2$ and $H^c = H_1 \vee H_2$ are nonisomorphic, connected graphs sharing the Laplacian spectrum,

$$s(G^c) = (12, 11, 11, 10, 9, 9, 9, 9, 8, 7, 7, 0)$$
$$= s(H^c). \qquad \square$$

9.27 Definition. If $s(G) = s(H)$, then graphs G and H are *isospectral*.

9.28 Example. Let G be the graph from Fig. 9.5 reproduced, for convenience, in Fig. 9.8. Because vertex u is a dominating vertex, $G \cong K_1 \vee H$, where $H \cong K_1 \oplus P_3$. Because $P_3 \cong K_1 \vee (K_1 \oplus K_1)$, we have $H \cong K_1 \oplus [K_1 \vee (K_1 \oplus K_1)]$ and

$$G \cong K_1 \vee (K_1 \oplus [K_1 \vee (K_1 \oplus K_1)]). \tag{69}$$

Figure 9.8

This formidable expression has a single virtue. It shows that G can be "decomposed" into joins and unions of isolated vertices. Because we know how joins and unions affect the spectra, $s(G)$ can be written down directly from Expression (69): $s(K_1) = (0)$, so $s(K_1 \oplus K_1) = (0, 0)$. From Corollary 9.25 (with $t = 1$ and $r = 2$), $s(P_3) = s(K_1 \vee [K_1 \oplus K_1]) = (3, 1, 0)$. So, $s(H) = (3, 1, 0, 0)$. A final application of Corollary 9.25 (with $t = 1$ and $r = 4$) yields $s(G) = (5, 4, 2, 1, 0)$, confirming Example 9.16. □

Two things about Example 9.28 are noteworthy: We were able to find the eigenvalues of $L(G)$ without looking at a matrix, much less computing its characteristic polynomial. And, in this case at least, the relationship of $s(G)$ to the structure of G is relatively clear.

9.29 Definition. A graph is *decomposable* if it can be expressed as joins and unions of isolated vertices.

Informally, G is decomposable if and only if it can be "constructed" from the raw material of isolated vertices, using only the tools of joins and unions.

9.30 Theorem. *The Laplacian spectrum of a decomposable graph consists entirely of integers.*

Proof. Suppose the eigenvalues of $L(G)$ and $L(H)$ are all integers. By Equation (67), the eigenvalues of $L(G \oplus H)$ are all integers; and by Corollary 9.25, $s(G \vee H)$ consists entirely of integers. Hence, the result follows from $s(K_1) = (0)$. ∎

9.31 Example. The Laplacian characteristic polynomial of P_4 is $\det(xI_4 - L(P_4)) = x(x-2)(x^2 - 4x + 2)$. Thus,

$$s(P_4) = (2 + \sqrt{2}, 2, 2 - \sqrt{2}, 0).$$

Because two of its eigenvalues are irrational, P_4 cannot be decomposable.

To take another example, suppose $G = C_6$, the graph G_1 from Example 9.26. Then $s(G) = (4, 3, 3, 1, 1, 0)$ consists entirely of integers. However, because G is connected, it is not a union of graphs; and because its complement (isomorphic to G_2 in Fig. 9.7) is connected, G is not a join of graphs. Thus, G cannot be decomposable. □

9.32 Theorem. *Let G be a graph on n vertices. Then G is decomposable if and only if it does not have an induced subgraph isomorphic to P_4.*

Proof. If P_4 is an induced subgraph of $H_1 \oplus H_2$, then P_4 is an induced subgraph of H_1 or of H_2. Because $P_4^c \cong P_4$, the same conclusion holds if P_4 is an induced subgraph of $H_1 \vee H_2 = (H_1^c \oplus H_2^c)^c$. Since P_4 is not a subgraph of K_1, decomposable graphs must be "P_4-free."

Conversely, suppose G does not have an induced subgraph isomorphic to P_4. If $G = K_1$ then, by definition, G is decomposable. Otherwise, by Theorem 1.19, one of G and G^c is disconnected. Therefore, G is either the union or the join of induced subgraphs having fewer than n vertices. Because these subgraphs are P_4-free, they are decomposable by induction. Since it is the union or join of decomposable subgraphs, G is decomposable. ■

9.33 Corollary. *Decomposable graphs are perfect.*

Proof. Theorem 9.32, Corollary 8.17, and the definition of perfect graph. ■

9.34 Corollary. *Threshold graphs are decomposable.*

Proof. Theorem 9.32 and Theorem 7.28. ■

Among the relationships between eigenvalues and graph structure is the following spectral characterization of threshold graphs.

9.35 Merris's Theorem.[8] *Let G be a graph with at least one edge. Then G is a threshold graph if and only if, apart from zeros, the Laplacian spectrum of G is equal to the conjugate of its degree sequence — that is, if and only if $s(G)$ is equivalent to $d(G)^*$.*

9.36 Example. If G is the graph illustrated in Figs. 9.5 and 9.8, then $d(G) = (4, 3, 2, 2, 1)$. From the division of $F(d(G))$ in Fig. 9.9, one sees that $\alpha(d(G)) = (4, 2) = \beta(d(G))$, so G is a threshold graph. By Theorem 9.35, $s(G) = d(G)^* = (5, 4, 2, 1, 0)$, confirming Examples 9.16 and 9.28. □

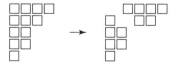

Figure 9.9

---

[8] R. Merris, Degree maximal graphs are Laplacian integral, *Linear Algebra Appl.* **199** (1994), 381–389.

Proof. Suppose G is a threshold graph with degree sequence $d(G) = \pi$. By Lemma 7.27, G has a unique nontrivial component C; C is a threshold graph with a dominating vertex u; and $H = C - u$ is a threshold graph on, say, r vertices. So, $C \cong K_1 \vee H$ and, because $G \cong C \oplus K_k^c$, $d(C)$ is equivalent to π. If $f(\pi) = 1$, then $d(C) = (r, 1, \ldots, 1)$, $\ell(\pi) = r + 1 = \pi_1^*$, $H = K_r^c$, and $s(H) = (0, 0, \ldots, 0)$. From Corollary 9.25 (with $t = 1$), $s(K_1 \vee H) = (r + 1, 1, \ldots, 1, 0)$, with $r - 1$ ones. Therefore, $s(C)$ is equal to the sequence obtained from π^* by appending a single zero, i.e., $s(C)$ is equivalent to π^*. Because $s(G)$ is obtained from $s(C)$ by appending k zeros, $s(G)$ is equivalent to π^*. This establishes the basis step for a proof by induction on $f(\pi)$.

If $f(\pi) > 1$, then $d(H) = (\pi_2 - 1, \ldots, \pi_{r+1} - 1)$. So, $F(d(H))$ is obtained from $F(\pi)$ by deleting the $2r$ boxes from its first row and column. By induction, $d(H)^*$ is equivalent to $s(H) = (\mu_1, \mu_2, \ldots, \mu_{r-1}, 0)$. By Corollary 9.25 (with $t = 1$ again), $s(C) = (r + 1, \mu_1 + 1, \ldots, \mu_{r-1} + 1, 0)$. So, $F(s(C))$ is obtained from $F(s(H)) = F(d(H))^t$ by adding a new first row and column containing $2r$ boxes, altogether. (See Fig. 9.10.) The result is precisely $F(\pi^*)$, the transpose of $F(\pi)$; that is, π^* is equivalent to $s(C)$ which is equivalent to $s(G)$.

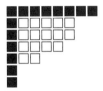

Figure 9.10. $F(s(C))$ from $F(s(H))$.

Conversely, let G be a graph with degree sequence $\pi = (\pi_1, \pi_2, \ldots, \pi_n)$ and Laplacian spectrum equivalent to π^*. If $l(\pi) = 2 = \pi_1^*$, then $G \cong K_2 \oplus K_{n-2}^c$ is a threshold graph.

Assume $p = \ell(\pi) > 2$ and let G' be the subgraph of G induced on its p vertices of positive degree. Without loss of generality, $L(G) = L(G') \oplus 0_{n-p}$, where $d_G(v_i) = \pi_i = d_{G'}(v_i)$, $1 \leq i \leq p$. Therefore, $\lambda_i(G') = \lambda_i(G) = \pi_i^*$, $1 \leq i \leq \pi_1$. Because $\lambda_1(G') = \pi_1^* = p$ we see, from Theorem 9.24, that $\lambda_{p-1}(G'^c) = 0$. From Corollary 9.21, G'^c is disconnected. Since a graph and its complement cannot both be disconnected, $0 \neq \lambda_{p-1}(G') = \pi_{p-1}^*$. It follows that $\pi_1 = p - 1$, meaning that v_1 is a dominating vertex of G'; that is, $G' \cong \{v_1\} \vee H$ for some graph H on $p - 1$ vertices. From Corollary 9.25,

$$\lambda_{i+1}(G') = \lambda_i(H) + 1, \quad 1 \leq i \leq p - 2. \tag{70}$$

Because $F(d(H))$ is obtained from $F(\pi)$ by removing its first row and column,

$$\pi_{i+1}^* = d_i^*(H) + 1, \quad 1 \leq i \leq p - 2. \tag{71}$$

Because $\lambda_{i+1}(G') = \pi^*_{i+1}$, $1 \leq i \leq p - 2$, it follows from Equations (70) and (71) that $\lambda_i(H) = d^*_i(H)$, $1 \leq i \leq p - 2$. So, by induction, H is a threshold graph. Therefore, $K_1 \vee H \cong G'$ is a threshold graph and $d(G')$ is a threshold partition. Since $d(G')$ is equivalent to $d(G)$, G is a threshold graph. ∎

When G is a tree, there is an interesting connection between its Laplacian spectrum and its Wiener index.

9.37 Theorem. *Let T be a tree with Laplacian spectrum $s(T) = (\lambda_1, \lambda_2, \ldots, \lambda_n)$. Then the Wiener index*

$$W(T) = n \sum_{i=1}^{n-1} \frac{1}{\lambda_i}. \tag{72}$$

While not difficult, the proof of Theorem 9.37 is beyond the scope of this book.[9]

9.38 Example. It is easily seen that the Wiener index of the 4-vertex path is $W(P_4) = (3 + 2 + 1) + (2 + 1) + 1 = 10$. (Confirm it.) From Example 9.31, $s(P_4) = (2 + \sqrt{2}, 2, 2 - \sqrt{2}, 0)$. Observing that

$$\frac{1}{2 + \sqrt{2}} + \frac{1}{2 - \sqrt{2}} = \frac{4}{4 - 2} = 2,$$

the right-hand side of Equation (72) is seen, in this case, to be $4 \times (2 + \frac{1}{2}) = 10$.

The Wiener index of the 4-vertex star is $W(S_4) = 9$. Since S_4 is a threshold graph, $s(S_4) = d(S_4)^* = (4, 1, 1, 0)$. (Check it.) In this case, the right-hand side of Equation (72) is $4 \times 2.25 = 9$. □

9.39 Example. Wiener's 1947 paper concerned paraffins — that is, molecules whose carbon skeletons are trees. One way to extend Wiener's index to graphs that are not trees is to define

$$W(G) = \sum d(u, v), \tag{73}$$

where the sum is over all $C(n, 2)$ 2-element subsets $\{u, v\}$ of $V(G)$, and $d(u, v)$ is the length of a shortest path in G from u to v. Using this definition, $W(G_1) = 71$ and $W(G_2) = 73$ for the isospectral[10] graphs in Fig. 9.11, proving that $W(G)$ is not any function of the Laplacian spectrum for graphs that are not trees.

---

[9] See R. Merris, An edge version of the matrix-tree theorem and the Wiener index, *Linear & Multilinear Algebra* **25** (1989), 292–296; R. Merris, The distance spectrum of a tree, *J. Graph Theory* **14** (1990), 365–369; or B. Mohar, Eigenvalues, diameter, and mean distance in graphs, *Graphs & Combinatorics* **7** (1991), 53–64.

[10] See C. Godsil, D. A. Holton, and B. McKay, The spectrum of a graph, *Lecture Notes in Mathematics* **662**, Springer-Verlag, New York, 1977, pp. 91–117.

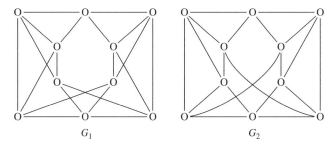

Figure 9.11

Recall that the *raison d'être* of a topological index is to predict chemical properties. It hardly seems possible that, from among all possible paths between two nonadjacent carbon atoms in an organic molecule, only a single shortest path could be relevant to the molecule's chemical properties. As we saw in the discussion leading up to Menger's Theorem, such is certainly not the case in graph models of social behavior. Neither is it the case for electrical networks.

Recall that if resistors R_1 and R_2 are connected in series, their combined resistance is $R = R_1 + R_2$. If the same resistors are connected in parallel, however, their effective resistance is given by $R = (R_1^{-1} + R_2^{-1})^{-1}$.

Suppose we view chemical graphs as electrical networks in which each edge is assumed (provisionally at least) to have resistance $R = 1$. Define the *resistance distance*, $\rho(u, v)$, to be the effective resistance between vertices u and v and let

$$W'(G) = \sum \rho(u, v), \qquad (74)$$

where the sum is over all $C(n, 2)$ 2-elements subsets of $V(G)$.

If T is a tree, the corresponding "network" has no parallel sections, so $\rho(u, v) = d(u, v)$ for all $u, v \in V(T)$. In this case, $W(T) = W'(T)$. Marine biologist D. J. Klein[11] and sociologist M. Altman[12] have argued that Equation (74), not Equation (73), is the natural generalization of the Wiener index to graphs that are not trees. Among Klein's many arguments is the fact that

$$W'(G) = n \sum_{i=1}^{n-1} \frac{1}{\lambda_i(G)}$$

for every connected graph G on n vertices. □

---

[11] D. J. Klein, Graph geometry, graph metrics, & Wiener, *Communications in Mathematical and in Computer Chemistry* **35** (1997), 7–27.
[12] M. Altman, Reinterpreting network measures for models of disease transmission, *Social Networks* **15** (1993), 1–17.

EXERCISES

1. Confirm Stanley's Theorem for the graph $G = \begin{smallmatrix} \circ - \circ \\ | \diagdown \\ \circ - \circ \end{smallmatrix}$.

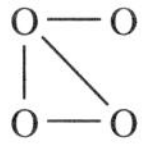

2. Compute the oriented vertex-edge incidence matrix $Q = Q(G)$ corresponding to the oriented graph

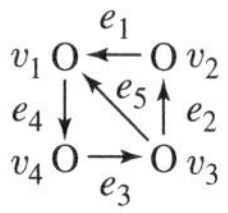

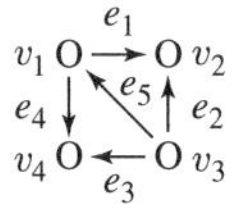

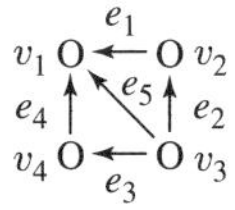

3. Confirm that QQ^t is the same for each part of Exercise 2.

4. Let G be the graph illustrated in Fig. 9.12.

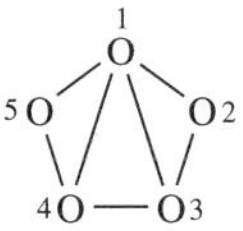

    Figure 9.12

    (a) Find $L(G)$.
    (b) Compute two different entries of the adjugate $L(G)^{\dagger}$.
    (c) Illustrate all $t(G)$ spanning trees of G.
    (d) Compute the number of acyclic orientations of G.

5. Compute the Laplacian spectrum of
    (a) C_4. (b) $K_4 - e \cong C_4 + e$. (c) S_4. (d) $S_4 + e$.

6. Prove the statement made in the text that G_1 and G_2 are isomorphic graphs if and only if $L(G_1)$ and $L(G_2)$ are permutation similar.

7. Confirm that the algebraic connectivity $a(G) \leq \kappa(G)$ for $G =$
    (a) C_4. (b) $C_4 + e$. (c) S_4. (d) P_4.

8. As in Chapter 6, Exercise 17, the line graph of G is the graph $G^{\#}$ whose vertex set is $V(G^{\#}) = E(G)$. The edges of $G^{\#}$ are those pairs of its vertices that are adjacent edges in G. If G is a *bipartite* graph, show that an orientation of G can always be chosen so that $Q(G)^t Q(G) = 2I_m + A(G^{\#})$, where $m = o(E(G))$.[13]

9. If T is a tree on $n \geq 3$ vertices, prove that $\lambda_{n-1}(T) \leq 1$.

---

[13] This result was used in I. Gutman, et al. [The high-energy band in the photoelectron spectrum of alkanes and its dependence on molecular structure, *J. Serb. Chem. Soc.* **64** (1999), 673–680] to express quantum-mechanical energy levels of certain paraffin molecules in terms of their Laplacian eigenvalues.

10 Confirm that the spanning tree number $t(C_6)$ equals 6
   (a) by showing that T is a spanning tree of $G = C_6$ if and only if $T = G - e$ for some fixed but arbitrary $e \in E(G)$.
   (b) using Equation (65) and Example 9.26.

11 Let $G = K_n$ for some fixed but arbitrary $n > 1$.
   (a) Show that $s(G) = (n, n, \ldots, n, 0)$.
   (b) Show that $t(G) = n^{n-2}$.

12 The tournament illustrated in Fig. 9.1 is not transitive. While team 1 beat team 3 and team 3 beat team 2, team 1 failed to beat team 2. An orientation of G is *transitive* if (u, w) is an oriented edge of G whenever (u, v) and (v, w) are among its oriented edges. A *comparability graph* is a graph that admits a transitive orientation.[14]

   (a) Show that a transitive orientation is acyclic.
   (b) Find an acyclic orientation of P_3 that is not transitive.
   (c) Show that every bipartite graph is a comparability graph.
   (d) Show that K_5 is a comparability graph.
   (e) Show that $G = $
   $$\begin{smallmatrix} & \circ - \circ & \\ & / \backslash & \\ \circ - \circ - \circ - \circ & \end{smallmatrix}$$
   is not a comparability graph.

13 Let $G = K_1 \vee (K_2 \oplus K_3)$ and $H = K_2^c \vee ([K_1 \vee K_2^c] \oplus K_1)$. Find
   (a) $s(G)$. (b) $s(G^c)$. (c) $s(G \oplus G^c)$. (d) $s(G \vee G^c)$.
   (e) $s(H)$. (f) $s(H^c)$. (g) $s(H \vee H^c)$.
   (h) Illustrate a pair of connected isospectral graphs that have different degree sequences.

14 Compute $t(G)$ if
   (a) $G = G_2$ in Fig. 9.7.
   (b) $G = H_1$ in Fig. 9.7.
   (c) $G = H_2$ in Fig. 9.7.

15 Let $G = K_{r,s}$, the complete bipartite graph.
   (a) Compute $s(G)$.
   (b) Compute $t(G)$.

16 Show that G is decomposable if and only if it can be constructed from the raw material of isolated vertices, using only the tools of complements and unions.

---

[14] It can be shown that every comparability graph is perfect.

17 Let G be a graph.

   (a) Prove that G is a threshold graph if and only if it can be expressed in the form $rK_1 \oplus [K_1 \vee (H \oplus tK_1)]$, where H is a connected threshold graph.

   (b) Prove that Chvátal's maximal nonhamiltonian graph $G_{n,r}$ is a threshold graph.

18 Let G be a connected, decomposable graph. Suppose $u, w \in V(G)$. Prove that the distance $d(u, w) \leq 2$.

19 Show that H_1 in Fig. 9.7 is not decomposable.

20 Show that

   (a) C_4 is a nonthreshold decomposable graph.

   (b) $2K_2$ is a nonthreshold decomposable graph.

21 Let G is a graph with n vertices and m edges. Prove that

$$\sum_{i=1}^{n} \lambda_i(G) = 2m.$$

22 It follows from a result of I. Schur[15] that the Laplacian spectrum of a graph majorizes its degree sequence. Confirm that $s(G) \succ d(G)$ for the graph

   (a) C_4. (b) P_4.
   (c) G_1 in Fig. 9.7. (d) G_2 in Fig. 9.7.
   (e) H_1 in Fig. 9.7. (f) H_2 in Fig. 9.7.

23 Let G be a graph with at least one edge. Recall (Chapter 7, Exercise 37) that $d(G)^*$, the conjugate of the degree sequence, majorizes $d(G)$. An unresolved conjecture[16] about the Laplacian spectrum is that $d(G)^* \succ s(G)$. Confirm this conjecture for the graph

   (a) C_4. (b) P_4.
   (c) G_1 in Fig. 9.7. (d) G_2 in Fig. 9.7.
   (e) H_1 in Fig. 9.7. (f) H_2 in Fig. 9.7.

24 Let G be a graph with vertex set $V(G) = \{v_1, v_2, \ldots, v_n\}$. Suppose $X = (x_1, x_2, \ldots, x_n)^t$ is an eigenvector of $L(G)$ affording eigenvalue λ. Assume $x_i = x_j$ for some $i \neq j$. Let G' be the graph obtained from G by deleting

---

[15] I. Schur, Über eine Klasse von Mittelbildungen mit Anwendungen die Determinantentheorie, *Sitzungsberichte Math. Gesellschaft Berlin* **22** (1923), 9–20.

[16] R. Grone and R. Merris, The Laplacian spectrum of a graph ii, *SIAM J. Discrete Math.* **7** (1994), 221–229.

or adding edge v_iv_j depending on whether or not it is an edge of G. Prove that X is an eigenvector of $L(G')$ affording λ.

25 Let $G = (V, E)$ be a bipartite graph with color classes X and Y. Suppose $d_G(v) = r$ for all $v \in X$ and $d_G(v) = s$ for all $v \in Y$. (Then G is said to be (r, s)-*semiregular*.) Prove that $\lambda = r + s$ is an eigenvalue of $L(G)$.

26 Let $e = uv$ be an edge of $G = (V, E)$.
  (a) Prove that the spanning tree number $t(G) = t(G - e) + t(G|e)$.
  (b) Use part (a) to evaluate $t(G)$ for the graph in Fig. 9.12.

27 Let $G = K_r \vee K_s^c$.
  (a) Compute $s(G)$.
  (b) Compute $t(G)$.

28 Let G be a graph with vertex set $V(G) = \{v_1, v_2, \ldots, v_n\}$. If $X = (x_1, x_2, \ldots, x_n)^t$, prove that
$$X^t L(G) X = \sum_{v_iv_j \in E(G)} (x_i - x_j)^2.$$

29 Recall that A_n is the unique connected antiregular graph on n vertices. Show that $s(A_n)$ consists of all but one of the integers $0, 1, \ldots, n$ and that the "missing" eigenvalue is $\lfloor (n+1)/2 \rfloor$.

30 Prove that the Laplacian spectrum of the star is $s(S_n) = (n, 1, \ldots, 1, 0)$.

31 Confirm that
  (a) $s(G_1) = (4, 3, 3, 1, 1, 0)$ for the graph $G_1 \cong C_6$ in Example 9.26.
  (b) $s(H_1) = (5, 3, 3, 2, 1, 0)$ for the graph H_1 in Example 9.26.

32 Let G be a graph with n vertices and $\xi(G)$ connected components. Prove that $n = \xi(G) + \text{rank } L(G)$.

33 Prove that the graphs in Fig. 9.13

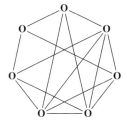
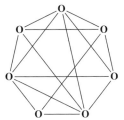

Figure 9.13

(a) are not isomorphic.

(b) are isospectral.

34 Let G be a connected graph with n vertices and m edges. Let $Q = Q(G)$ be an oriented vertex-edge incidence matrix for G.

(a) Prove that the $n - 1$ columns of Q corresponding to a spanning tree of G are linearly independent.

(b) Prove that the columns of Q corresponding to the edges of a cycle of G are linearly dependent.

35 The purpose of this exercise is to explore a little more deeply the innocent observation made in Exercise 34(b). Suppose $G = (V, E)$, where $V = \{v_1, v_2, \ldots, v_n\}$ and $E = \{e_1, e_2, \ldots, e_m\}$. Let $Q = Q(G)$ be an oriented vertex-edge incidence matrix for G. Denote the kernel of Q by W; that is, $W = \{m \times 1$ matrices Z such that $QZ = 0\}$.

(a) Prove that rank $Q = n - \xi(G)$.

(b) Prove that $\dim(W) = m - n + \xi(G)$.

(c) Suppose $f : E(G) \to \mathbb{R}$ is a real-valued function of $E(G)$. Let X_f be the $m \times 1$ matrix whose jth entry is $f(e_j)$. Prove that QX_f is the $n \times 1$ matrix whose ith entry is

$$\sum_{\text{head}(e)=v_i} f(e) - \sum_{\text{tail}(e)=v_i} f(e).$$

(d) A *flow* in the oriented graph G is a real-valued function f of its edges such that

$$\sum_{\text{head}(e)=v} f(e) = \sum_{\text{tail}(e)=v} f(e),$$

for every $v \in V$.[17] Prove that $f : E(G) \to \mathbb{R}$ is a flow if and only if $X_f \in W$.

(e) Suppose $C = \langle 1, v_2, \ldots, v_r \rangle$ is a cycle in G. Let $v_{r+1} = v_1$. Define the signed characteristic function $f_C : E(G) \to \mathbb{R}$ by $f_C(e) = 0$ if e is not an edge of C; $f_C(e) = 1$ if e is an edge of C with a compatible orientation, that is, if $e = (v_i, v_{i+1})$ for some i; and $f_C(e) = -1$ if $e = (v_{i+1}, v_i)$ for some i. Prove that $X_{f_C} \in W$.

(f) Suppose G is connected. Let T be a fixed but arbitrary spanning tree of G and suppose $e \in E(G) \setminus E(T)$. Prove that the spanning subgraph $H = (V, E(T) \cup \{e\})$ of G contains a unique cycle $C(T, e)$.

(g) Prove that $\{X_{f_{C(T,e)}} : e \in E(G) \setminus E(T)\}$ is a basis for W.[18]

---

[17] Among Kirchhoff's "Laws" is the fact that current in an electrical network behaves like a flow in the underlying graph.

[18] The kernel of Q is sometimes called the *cycle space* of G. These ideas are developed more fully in [N. Biggs, *Algebraic Graph Theory*, 2nd ed., Cambridge University Press, Cambridge, England,

36 Confirm Equation (72)

(a) for the star S_n.

(b) for the tree $T = \text{o—o—o—o}$ with an o attached above the second o.

37 Prove or disprove that any $n - 1$ edges of K_n comprise the edge set of some spanning tree of K_n.

38 It was conjectured by A. Gyárfás and J. Lehel that if T_k is a fixed but arbitrary tree on k vertices, $2 \leq k \leq n$, then isomorphic copies of $T_2, T_3, \ldots, T_n$ can be found in K_n where no two copies share an edge — that is, where

$$E(K_n) = \bigcup_{k \geq 2} E(T_k).$$

(a) Confirm that $o(E(K_n)) = \sum_{k \geq 2} o(E(T_k))$ for any such collection of trees.

(b) Prove that the edge set of K_n can be partitioned as the disjoint union of the edge sets of stars, S_k, $2 \leq k \leq n$.

(c) Prove that the edge set of K_7 can be partitioned as the disjoint union of the edge sets of paths, P_k, $2 \leq k \leq 7$.

39 (Roy and Gallai) Suppose $\chi(G) < k$. Prove that G has an orientation in which no directed path has length k or more.

40 Prove that every tournament (oriented complete graph) has a spanning directed path.

41 Let $G = A_n$ be the unique, connected, antiregular graph on n vertices. If T is a tree on n vertices, prove that T is isomorphic to a spanning tree of G.

42 Let G and H be the graphs in Fig. 9.14.

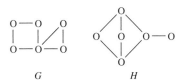

Figure 9.14

(a) Show that $\det(xI_6 - L(G)) = x(x - 2)(x - 3)^2(x^2 - 6x + 4)$.

(b) Show that $\det(xI_6 - L(H)) = x(x - 2)(x - 3)^2(x^2 - 6x + 4)$.

1993]. Also see [N. Biggs, Algebraic potential theory on graphs, *Bull. London Math. Soc.* **29** (1997), 641–682].

(c) Show that G^c is isomorphic to a graph that can be obtained from H by the addition of a single well-chosen edge.

(d) Show that H^c is isomorphic to a graph that can be obtained from G by the addition of a single well-chosen edge.

(e) Compute the Laplacian spectrum, $s(G^c) = s(H^c)$.

43 Denote the characteristic polynomial of $L(G)$ by

$$\det(xI_n - L(G)) = x^n - b_1 x^{n-1} + b_2 x^{n-2} - \cdots + (-1)^{n-1} b_{n-1} x.$$

Let $\mathscr{F}_t$ be the set of all t-edged spanning forests of G. If $F \in \mathscr{F}_t$, denote by $\pi(F)$ the product of the numbers of vertices in the $n - t$ components of F. It was shown by A. K. Kel'mans and V. M. Chelnokov that $b_t = \sum \pi(F)$ as F ranges over $\mathscr{F}_t$.

(a) Use the result of Kel'mans and Chelnokov to prove that $b_t \geq (t+1)c_t$, $1 \leq t < n$, where $(-1)^t c_t$ is the coefficient of x^{n-t} in the chromatic polynomial $p(G, x)$. (*Hint*: Whitney's Broken Cycle Theorem.)

(b) Prove that $b_{n-1} = nc_{n-1}$ if and only if G is a tree.

10

Edge Colorings

It is common practice, among aficionados of internet chat, to use aliases. While two such individuals might be in frequent electronic communication, they could easily fail to recognize one another when meeting face-to-face—say, at a COMDEX convention. Let's say two people are *e-quainted* if, during the past year, they participated together in one or more of the same chat room discussions. Otherwise, call them *e-stranged*. Then any two people attending the convention's introductory cocktail party will be either e-quainted or e-stranged.

Would it surprise you to learn that every group of six cocktail party guests must contain three mutual e-quaintances or three mutual e-strangers?[1] To see why, let Arwen be one of the six. Then the other five members of the group fall into one of two categories, according to whether or not they are e-quainted with Arwen, and one of these categories must contain three (or more) people. Suppose Balthazar, Carse, and Danilo belong to the category of people e-stranged from Arwen. If these three characters are mutual e-quaintances, we are finished. Otherwise, some two of them are e-stranged and these two, together with Arwen, are mutually e-stranged.

If the category of people e-quainted with Arwen contains three people and if these three are mutually e-stranged, we are finished. Otherwise, some two of them are e-quainted and these two, together with Arwen, are mutually e-quainted.

Let's model the situation using graphs. Begin by identifying some group of six guests with the vertices of K_6. If guests i and j are e-quainted, color edge $v_i v_j$ black; if they are not, color it white. From this perspective, our observation is that the resulting figure contains a black triangle or a white triangle.

There is another way to look at this model. Let G be a graph on six vertices. Visualize G as a spanning subgraph of some fixed illustration of K_6. Then "white out" the edges that do not belong to G. This process leads to a natural one-to-one correspondence between the different black–white edge colorings of K_6 and the different graphs on 6 vertices. In this correspondence a black triangle in the edge coloring represents a clique in the corresponding graph, and a white triangle represents an independent set. Evidently, $\omega(G) \geq 3$ or $\alpha(G) \geq 3$, for every graph

---

[1] An unqualified "or" should be understood in the inclusive sense, as in "*A* or *B* or both."

G on 6 vertices. Moreover, because $\omega(C_5) = 2 = \alpha(C_5)$, 6 is the smallest number of vertices sufficient to guarantee this *Ramsey property*.[2]

It is a consequence of Theorem 10.3 (below) that, for every pair of positive integers s and t, there exists an integer N such that every graph on N vertices contains an s-vertex clique or an independent set of t vertices. If G is a graph on $n > N$ vertices, let H be the subgraph of G induced on some fixed but arbitrary collection of N of its vertices. Because $\omega(G) \geq \omega(H)$ and $\alpha(G) \geq \alpha(H)$, G itself must satisfy the Ramsey property for s and t.

10.1 Definition. Let s and t be positive integers. The *Ramsey number*, $r(s, t)$, is the smallest integer such that every graph on $r(s, t)$ vertices has an induced subgraph isomorphic to K_s or an induced subgraph isomorphic to K_t^c.

In view of Definition 10.1, our cocktail party observation can be stated in the form $r(3, 3) = 6$.

10.2 Example. Let s and t be positive integers. Since $G \to G^c$ maps the set of graphs on n vertices to itself, and because $\omega(G) = \alpha(G^c)$, Ramsey numbers are symmetric; that is,

$$r(s, t) = r(t, s).$$

Because any 1-vertex set is independent, $r(s, 1) = 1$. Since $\alpha(G) = 2$ for all s-vertex graphs $G \neq K_s$, $r(s, 2) \leq s$. Because $\omega(K_n) < s$ and $\alpha(K_n) = 1 < 2$, for all $n < s$, $r(s, 2) \geq s$. So,

$$r(s, 1) = 1 = r(1, t),$$
$$r(s, 2) = s, \quad \text{and} \quad r(2, t) = t,$$

for all integers $s, t \geq 1$.

These relatively easy observations can be misleading. Finding the exact value, say, of $r(9, 10)$ seems to be out of reach of current techniques. Indeed, every presently known Ramsey number can be deduced from the easy observations and the table in Fig. 10.1. □

	1	2	3	4	5	6	7	8	9
1	1	1	1	1	1	1	1	1	1
2	1	2	3	4	5	6	7	8	9
3	1	3	6	9	14	18	23	28	36
4	1	4	9	18	25	?	?	?	?
5	1	5	14	25	?	?	?	?	?
6	1	6	18	?	?	?	?	?	?

Figure 10.1. Table of Ramsey numbers, $r(s, t) = r(t, s)$

---
[2] After Frank Ramsey (1902–1930).

It is hard to look very long at Fig. 10.1 and not begin searching for some Pascal-like recurrence that would identify the missing entries. There is a result along these lines, but it is not the expected identity.

10.3 Theorem. *Suppose $s, t \geq 2$. If G is a graph on $n = r(s, t - 1) + r(s - 1, t)$ vertices, then $\omega(G) \geq s$ or $\alpha(G) \geq t$; that is, the Ramsey number*

$$r(s, t) \leq r(s, t - 1) + r(s - 1, t). \tag{75}$$

Until now, we have been dealing with Ramsey numbers based upon a promise that $r(s, t)$ exists. Example 10.2 and Theorem 10.3 give us the inductive means to fulfill that promise.

Proof. Let G be a fixed but arbitrary graph on $n = r(s, t - 1) + r(s - 1, t)$ vertices. We desire to show that $\omega(G) \geq s$ or $\alpha(G) \geq t$. Our approach is a variation on the theme of cocktail party guests. Begin by choosing a vertex $w \in V(G)$.

Case 1: If $d_G(w) \geq r(s - 1, t)$, let H be the subgraph of G induced on $N_G(w)$, the neighbors of w. If $\alpha(H) \geq t$, then, because $\alpha(G) \geq \alpha(H)$, the proof is finished. Otherwise, from the definition of $r(s - 1, t)$, H contains an $(s - 1)$-vertex clique which, together with w, produces an s-vertex clique in G.

Case 2: If $d_G(w) < r(s - 1, t)$, let H be the graph that remains after w and all of its neighbors have been deleted from G. Then $o(V(H)) = n - (d_G(w) + 1) \geq n - r(s - 1, t) = r(s, t - 1)$. If $\omega(H) \geq s$, the proof is finished. Otherwise, H contains an independent set of $t - 1$ vertices which, together with w, is a t-vertex independent set in G. ■

10.4 Corollary. *For any positive integers s and t, the Ramsey number $r(s, t) \leq C(s + t - 2, t - 1)$.*

Proof. If $t = 1$, then $r(s, t) = r(s, 1) = 1 = C(s - 1, 0) = C(s + t - 2, t - 1)$. If $t = 2$, then $r(s, t) = r(s, 2) = s = C(s, 1) = C(s + t - 2, t - 1)$. By symmetry, the same arguments apply when $s \leq 2$. We proceed by induction on $k = s + t$.

Suppose $s, t \geq 3$ and assume the result is true for all $k < s + t$. Then, by Theorem 10.3,

$$r(s, t) \leq r(s, t - 1) + r(s - 1, t)$$
$$\leq C(s + t - 3, t - 2) + C(s + t - 3, t - 1)$$
$$= C(s + t - 2, t - 1).$$

■

We turn now to lower bounds.

10.5 Theorem. *The Ramsey number $r(s, t) \geq (s - 1)(t - 1) + 1$.*

Proof. If $s = 1$ or $t = 1$, then $r(s, t) = 1 = (s - 1)(t - 1) + 1$. So, assume $s, t \geq 2$ and let $n = (s - 1)(t - 1)$. Arrange the vertices of K_n in a rectangular array of $s-1$ rows and $t-1$ columns. If vertices u and v lie in the same row of the array, color edge uv white. Otherwise, color it black. Let G be the graph on n vertices corresponding to the black edges. (Then $G \cong K^c_{s-1} \vee K^c_{s-1} \vee \cdots \vee K^c_{s-1}$, the join of $t-1$ copies of K^c_{s-1}.)

If W is any set of s vertices of G, then some two of them must come from the same row of the array. Because these vertices are not adjacent, the induced subgraph $G[W] \not\cong K_s$. If U is any set of t vertices of G, then some two of them must come from the same column of the array. Since two vertices cannot lie in the same row if they come from the same column, these two vertices are adjacent; that is, $G[U] \not\cong K^c_t$. Therefore, $\omega(G) < s$ and $\alpha(G) < t$. Because G does not satisfy the Ramsey property for s and t, we have $r(s, t) > n$. ∎

10.6 Example. From Example 10.2 and our cocktail party observation, $r(2, 4) = 4$ and $r(3, 3) = 6$. By Theorem 10.3, $r(3, 4) \leq r(3, 3) + r(2, 4) = 10$. If $s = 3$ and $t = 4$, then $(s-1)(t-1) = 6$ and, by Theorem 10.5, $r(3, 4) \geq 7$. From Fig. 10.1, the actual value is $r(3, 4) = 9$.

Figure 10.1 gives $r(5, 4) = r(4, 5) = 25$, which, together with Theorem 10.3, yields $r(5, 5) \leq 50$. By Theorem 10.5, $r(5, 5) \geq 4 \times 4 + 1 = 17$. So, $17 \leq r(5, 5) \leq 50$. The bounds

$$43 \leq r(5, 5) \leq 49$$

are not so easy to obtain.[3] □

10.7 Erdös's Theorem[4]. *If $s \geq 2$, then $r(s, s) \geq \sqrt{2}^s$.*

Because $\sqrt{2}^5 \doteq 5.7$, Theorem 10.7 yields $r(5, 5) \geq \lceil 5.7 \rceil = 6$, not nearly as good as the lower bound $r(5, 5) \geq 17$ from Theorem 10.5. On the other hand, $(s-1)(s-1) + 1 = s^2 - 2s + 2$ bounds $r(s, s)$ by a polynomial in s, whereas $r(s, s) \geq \sqrt{2}^s$ shows that $r(s, s)$ grows exponentially in s.

Proof. Because $r(2, 2) = 2$, we may assume $s \geq 3$. Let $n = \lfloor 2^{s/2} \rfloor$, the integer part of $\sqrt{2}^s$. Let $W = \{1, 2, \ldots, n\}$ and denote by $\mathscr{G}_n$ the set of all (different) graphs with vertex set W. The idea of the proof is to show it cannot be true that $\omega(G) \geq s$ or $\alpha(G) \geq s$, for all $G \in \mathscr{G}_n$.

Consider a fixed but arbitrary s-element set $S \subset W$. For how many graphs in $\mathscr{G}_n$ is S a clique? Because $V(G) = W$ for all $G \in \mathscr{G}_n$, choosing G from $\mathscr{G}_n$ is equivalent to choosing $E(G)$, and this is equivalent to making a sequence of $C(n, 2)$ decisions, each involving 2 choices. For S to be a clique in G, $C(s, 2)$ of those decisions have already been made. By the Fundamental Counting Principle, the number of ways to make the remaining decisions is $2^{C(n,2)-C(s,2)}$. Therefore, S is a clique in $2^{C(n,2)}/2^{C(s,2)}$ of the $2^{C(n,2)}$ graphs in $\mathscr{G}_n$.

---
[3] See B.D. McKay and S.P. Radziszowski, Subgraph counting identities and Ramsey numbers, *J. Combinatorial Theory* B **69** (1997), 193–209.
[4] P. Erdös, Some remarks on the theory of graphs, *Bull. Amer. Math. Soc.* **53** (1947), 292–294.

Let $k = o(\{G \in \mathscr{G}_n : \omega(G) \geq s\})$. Because there are $C(n, s)$ ways to choose S, we have

$$k \leq C(n, s) 2^{C(n,2)} / 2^{C(s,2)}.$$

(Since K_n, for example, has been counted $C(n, s)$ times on the right-hand side of this expression, it is correctly phrased as an inequality.) Because $C(n, s) = n^{(s)}/s! < n^s/s!$, it follows that

$$k < \frac{n^s 2^{C(n,2)}}{s! 2^{C(s,2)}}. \tag{76}$$

Since $n \leq 2^{s/2}$,

$$n^s \leq (2^{s/2})^s$$
$$= 2^{C(s,2)+(s/2)}$$
$$= 2^{s/2} 2^{C(s,2)}.$$

Thus, from Inequality (76),

$$k < (2^{s/2}/s!) 2^{C(n,2)},$$
$$< \tfrac{1}{2} 2^{C(n,2)},$$

for all $s \geq 3$. Therefore, fewer than half of the graphs in $\mathscr{G}_n$ satisfy $\omega(G) \geq s$. Because $\mathscr{G}_n = \{G^c : G \in \mathscr{G}_n\}$, the same argument shows that $\omega(G^c) = \alpha(G) \geq s$ for fewer than half of the graphs in $\mathscr{G}_n$. Therefore,

$$o(\{G \in \mathscr{G}_n : \omega(G) \geq s \text{ or } \alpha(G) \geq s\})$$
$$= o(\{G \in \mathscr{G}_n : \omega(G) \geq s\} \cup \{G \in \mathscr{G}_n : \alpha(G) \geq s\})$$
$$\leq o(\{G \in \mathscr{G}_n : \omega(G) \geq s\}) + o(\{G \in \mathscr{G}_n : \alpha(G) \geq s\})$$
$$< o(\mathscr{G}_n)/2 + o(\mathscr{G}_n)/2$$
$$= o(\mathscr{G}_n).$$

So, it is not true that $\omega(G) \geq s$ or $\alpha(G) \geq s$ for all $G \in \mathscr{G}_n$. ∎

We proceed now to another application of the natural correspondence between graphs on n vertices and black–white edge colorings of K_n.

10.8 Definition. Suppose n is a fixed but arbitrary positive integer. Let $g(n, m)$ be the number of nonisomorphic graphs having n vertices and m edges. Then the *graph generating function*

$$f_n(x) = \sum_{m \geq 0} g(n, m) x^m.$$

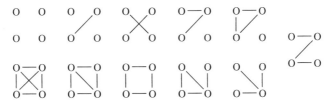

Figure 10.2

10.9 Example. The 11 nonisomorphic graphs on 4 vertices are illustrated in Fig. 10.2. From this figure, it is easy to write down

$$f_4(x) = 1 + x + 2x^2 + 3x^3 + 2x^4 + x^5 + x^6. \tag{77}$$

(Confirm that $\sum_{m \geq 0} g(4, m) = f_4(1) = 11$.) □

Evidently, the degree of $f_n(x)$ is $C(n, 2)$. Because K_n is the unique graph with n vertices and $C(n, 2)$ edges, $f_n(x)$ is a monic polynomial. Moreover, because $G_1 \cong G_2$ if and only if $G_1^c \cong G_2^c$, the coefficients of f_n are symmetric in the sense that

$$g(n, m) = g(n, C(n, 2) - m), \qquad 0 \leq m \leq C(n, 2). \tag{78}$$

With an illustration, comparable to Fig. 10.2, for the nonisomorphic graphs on 5 vertices, it would be just as easy to write down

$$f_5(x) = 1 + x + 2x^2 + 4x^3 + 6x^4 + 6x^5 + 6x^6 \\ + 4x^7 + 2x^8 + x^9 + x^{10}. \tag{79}$$

(Before reading on, compute $f_5(1) = \sum_{m \geq 0} g(5, m)$, the number of nonisomorphic graphs on 5 vertices.)

Suppose it were *your* job to illustrate the 34 nonisomorphic graphs on 5 vertices. In that case, it could be very useful to have $f_5(x)$ at your disposal — to know, for example, that exactly six of the graphs have four edges. What's needed is a recipe for computing $f_n(x)$, one that does not depend on already having a list of the graphs. Such a recipe is given in Theorem 10.16 (below); an application of the recipe to derive $f_5(x)$ is carried out in Example 10.18.

Let V be a fixed but arbitrary n-element set. Recall that any graph on n vertices is isomorphic to a graph with vertex set V. Moreover, from the definition of isomorphism, $G_1 = (V, E_1)$ is isomorphic to $G_2 = (V, E_2)$ if and only if there is a permutation p of V (i.e., a one-to-one function $p: V \to V$) such that

$$\{x, y\} \in E_1 \quad \text{if and only if} \quad \{p(x), p(y)\} \in E_2.$$

10.10 Definition. Suppose V is an n-element set. Let S_V be the family of all $n!$ permutations of V. For each $p \in S_V$, the permutation of $V^{(2)}$ *induced* by p is the function $\tilde{p}: V^{(2)} \to V^{(2)}$ defined by

$$\tilde{p}(\{x, y\}) = \{p(x), p(y)\}.$$

The set of all these induced permutations is the so-called *pair group*, $S_V^{(2)} = \{\tilde{p}: p \in S_V\}$.

Suppose $V = \{1, 2, \ldots, n\}$. Let $G_1 = (V, E_1)$ and $G_2 = (V, E_2)$ be graphs on vertex set V. Think of E_i as the set of black edges in a black–white coloring of $V^{(2)} = E(K_n)$, $i = 1, 2$. Then $G_1 = (V, E_1)$ is isomorphic to $G_2 = (V, E_2)$ if and only if there is an induced permutation $\tilde{p} \in S_V^{(2)}$ that maps black edges to black edges and white edges to white edges — that is, such that

$$e \in E_1 \quad \text{if and only if} \quad \tilde{p}(e) \in E_2. \tag{80}$$

10.11 Example. If $V = \{1, 2, 3, 4\}$, then S_V is usually denoted S_4 (not to be confused with the star on 4 vertices). If we number the elements of $V^{(2)}$ in "dictionary order,"

$$\mathbf{1} = \{1, 2\}, \quad \mathbf{2} = \{1, 3\}, \quad \mathbf{3} = \{1, 4\}, \quad \mathbf{4} = \{2, 3\}, \quad \mathbf{5} = \{2, 4\}, \quad \mathbf{6} = \{3, 4\},$$

then each $\tilde{p} \in S_V^{(2)}$ can be identified with a permutation in S_6. Suppose, for example, that $p = (143) \in S_V$. Then the disjoint cycle factorization of $\tilde{p} \in S_V^{(2)}$ is computed as follows:

$$\tilde{p}(\mathbf{1}) = \tilde{p}(\{1, 2\}) = \{p(1), p(2)\} = \{4, 2\} = \mathbf{5}$$
$$\tilde{p}(\mathbf{5}) = \tilde{p}(\{2, 4\}) = \{p(2), p(4)\} = \{2, 3\} = \mathbf{4}$$
$$\tilde{p}(\mathbf{4}) = \tilde{p}(\{2, 3\}) = \{p(2), p(3)\} = \{2, 1\} = \mathbf{1},$$

so, $(\mathbf{154})$ is one of the cycles in the disjoint cycle factorization of $\tilde{p}$. Continuing,

$$\tilde{p}(\mathbf{2}) = \tilde{p}(\{1, 3\}) = \{p(1), p(3)\} = \{4, 1\} = \mathbf{3}$$
$$\tilde{p}(\mathbf{3}) = \tilde{p}(\{1, 4\}) = \{p(1), p(4)\} = \{4, 3\} = \mathbf{6}$$
$$\tilde{p}(\mathbf{6}) = \tilde{p}(\{3, 4\}) = \{p(3), p(4)\} = \{1, 3\} = \mathbf{2}.$$

Thus, $\tilde{p} = (\mathbf{154})(\mathbf{236})$. Similar computations lead to the table in Fig. 10.3, where each $\tilde{p} \in S_V^{(2)}$ has been identified with an element of S_6. Note that $o(S_V^{(2)}) = o(S_V) = 4! = 24$, a small fraction of the $6! = 720$ permutations in S_6. □

p	$\tilde{p}$	p	$\tilde{p}$	p	$\tilde{p}$
id.	**id.**	(13) (24)	(**16**) (**34**)	(123)	(**142**) (**356**)
(12)	(**24**) (**35**)	(14) (23)	(**16**) (**25**)	(124)	(**153**) (**246**)
(13)	(**14**) (**36**)	(1234)	(**1463**) (**25**)	(132)	(**124**) (**365**)
(14)	(**15**) (**26**)	(1243)	(**1562**) (**34**)	(134)	(**145**) (**263**)
(23)	(**12**) (**56**)	(1324)	(**16**) (**2453**)	(142)	(**135**) (**264**)
(24)	(**13**) (**46**)	(1342)	(**1265**) (**34**)	(143)	(**154**) (**236**)
(34)	(**23**) (**45**)	(1423)	(**16**) (**2354**)	(234)	(**123**) (**465**)
(12) (34)	(**25**) (**34**)	(1432)	(**1364**) (**25**)	(243)	(**132**) (**456**)

Figure 10.3. S_V and $S_V^{(2)} \subset S_6$ for $V = \{1, 2, 3, 4\}$

10.12 Definition. Suppose V is an n-element set. Let $s_1, s_2, \ldots, s_k$ be $k = C(n, 2)$ independent variables. Then the *cycle index polynomial* of $S_V^{(2)}$ is

$$Z_n(s_1, s_2, \ldots, s_k) = \frac{1}{n!} \sum_{p \in S_V} s_1^{c_1(p)} s_2^{c_2(p)} \cdots s_k^{c_k(p)},$$

where $c_i(p)$ is the number of cycles of length i in the disjoint cycle factorization, not of p, but of $\tilde{p}$.

From Fig. 10.3 (where cycles of length 1 do not explicitly appear), it is easy to produce the cycle index polynomial

$$Z_4(s_1, s_2, \ldots, s_6) = (s_1^6 + 9s_1^2 s_2^2 + 6s_2 s_4 + 8s_3^2)/24. \tag{81}$$

Note the absence of s_5 and s_6 from the right-hand side of Equation (81).

It turns out that, to find the cycle index polynomial, it is not necessary to compute the disjoint cycle factorization of $\tilde{p}$ for each and every $p \in S_V$.

10.13 Lemma. *Let $\tilde{p}$ and $\tilde{q}$ be the elements of $S_V^{(2)}$ induced by the permutations p and q of S_V, respectively. If p and q have the same cycle structure, then $\tilde{p}$ and $\tilde{q}$ have the same cycle structure.*

Proof. Let $p \in S_V$. Fix $i, j \in V$. Let p' be the permutation obtained by interchanging the positions of i and j in the disjoint cycle factorization of p. Then p' and p have the same cycle structure. Moreover, $\tilde{p}'$, can be obtained by switching the positions of $\mathbf{r} = \{i, k\}$ and $\mathbf{t} = \{j, k\}$ in the disjoint cycle factorization of $\tilde{p}$ for every k different from i and j. Therefore, $\tilde{p}'$ and $\tilde{p}$ have the same cycle structure. Because p and q have the same cycle structure if and only if q can be obtained from p by a sequence of such interchanges, the proof is complete. ∎

10.14 Example. To see that the converse of Lemma 10.13 is false, let $V = \{1, 2, 3, 4\}$, $p = (12)(34)$, and $q = (34)$. From Fig. 10.3, $\tilde{p}$ and $\tilde{q}$ share the cycle structure (**ij**) (**xy**). In particular, both contribute 1 to the coefficient of $s_1^2 s_2^2$ in the cycle index polynomial $Z_4(s_1, s_2, \ldots, s_6)$. □

10.15 Example. Let $V = \{1, 2, 3, 4, 5\}$. Number the $C(5, 2) = 10$ elements of $V^{(2)}$ in dictionary order:

$$\mathbf{1} = \{1, 2\}, \quad \mathbf{2} = \{1, 3\}, \quad \mathbf{3} = \{1, 4\}, \quad \mathbf{4} = \{1, 5\}, \quad \mathbf{5} = \{2, 3\},$$
$$\mathbf{6} = \{2, 4\}, \quad \mathbf{7} = \{2, 5\}, \quad \mathbf{8} = \{3, 4\}, \quad \mathbf{9} = \{3, 5\}, \text{ and } \mathbf{0} = \{4, 5\}.$$

Of the $5! = 120$ elements of S_V, 10 are *transpositions*—that is, permutations having cycle structure $(i\,j)$. If, for example $q = (12)$, then, as in Example 10.11, $\tilde{q} = (\mathbf{25})(\mathbf{36})(\mathbf{47})$. (Confirm it.) Thus, the transpositions in S_V contribute a total of $10s_1^4 s_2^3$ to the cycle index polynomial $Z_5 = Z_5(s_1, s_2, \ldots, s_{10})$. Apart from the identity that contributes s_1^{10} to Z_5, representative permutations $p \in S_V$ and the corresponding $\tilde{p} \in S_V^{(2)}$ are given in the table in Fig. 10.4. The column headed "#" gives the number of permutations in S_V having the cycle type represented by p. From this figure, we obtain

$$Z_5(s_1, s_2, \ldots, s_{10})$$
$$= (s_1^{10} + 10 s_1^4 s_2^3 + 15 s_1^2 s_2^4 + 20 s_1 s_3^3 + 20 s_1 s_3 s_6 + 30 s_2 s_4^2 + 24 s_5^2)/120. \quad (82)$$

Note, from Equations (81) and (82), that $Z_4(1, 1, \ldots, 1) = 1 = Z_5(1, 1, \ldots, 1)$, reflecting the fact that $Z_n(s_1, s_2, \ldots, s_k)$ is the mean of $n!$ monomials. □

p	$\tilde{p}$	#	p	$\tilde{p}$	#
(12)	(**25**) (**36**) (**47**)	10	(123) (45)	(**152**) (**378469**)	20
(12) (34)	(**26**) (**35**) (**47**) (**90**)	15	(1234)	(**1583**) (**26**) (**4790**)	30
(123)	(**152**) (**368**) (**479**)	20	(12345)	(**15804**) (**26937**)	24

Figure 10.4. p and $\tilde{p}$ when $V = \{1, 2, 3, 4, 5\}$

We come at last to the result that makes cycle index polynomials worth knowing about.

10.16 Theorem[5]**.** *The graph generating function $f_n(x)$ is obtained by substituting $s_i = 1 + x^i$, $1 \le i \le k = C(n, 2)$, in the cycle index polynomial $Z_n(s_1, s_2, \ldots, s_k)$; that is,*

$$f_n(x) = Z_n(1 + x, 1 + x^2, \ldots, 1 + x^k). \quad (83)$$

The proof of Theorem 10.16 is beyond the scope of this book.

---

[5] Theorem 10.16 is the restriction to graph enumeration of a much more general result of G. Pólya [Kombinatorische Anzahlbestimmungen für Gruppen, Graphen und chemische Verbindungen, *Acta Mathematica*, **68** (1937), 145–254] and J. H. Redfield [The theory of group reduced distributions, *Amer. J. Math.* **49** (1927), 433–455]. "Pólya's Theorem" can be found in many combinatorics books, for example, R. Merris, *Combinatorics*, PWS & Brooks/Cole, Pacific Grove, CA, 1996.

10.17 Example. From Theorem 10.16 and Equation (81), we have

$$f_4(x) = \frac{1}{24}[(1+x)^6 + 9(1+x)^2(1+x^2)^2 + 6(1+x^2)(1+x^4) + 8(1+x^3)^2]$$

$$= \frac{1}{24}[(1 + 6x + 15x^2 + 20x^3 + 15x^4 + 6x^5 + x^6)$$
$$+ 9(1 + 2x + 3x^2 + 4x^3 + 3x^4 + 2x^5 + x^6)$$
$$+ 6(1 + x^2 + x^4 + x^6) + 8(1 + 2x^3 + x^6)]$$
$$= 1 + x + 2x^2 + 3x^3 + 2x^4 + x^5 + x^6,$$

precisely Equation (77). □

10.18 Example. From Equation (82), we obtain

$$Z_5(s_1, s_2, \ldots, s_{10})$$
$$= (s_1^{10} + 10s_1^4 s_2^3 + 15s_1^2 s_2^4 + 20s_1 s_3^3 + 20s_1 s_3 s_6 + 30s_2 s_4^2 + 24s_5^2)/120. \quad (84)$$

Let's use this expression and Theorem 10.16 to compute $f_5(x)$. From Equation (78), it suffices to evaluate the coefficient of x^m, $0 \le m \le 5$. From the binomial theorem, we have

$$(1+x)^{10} = 1 + 10x + 45x^2 + 120x^3 + 210x^4 + 252x^5 + \cdots.$$

Substituting $s_1 = 1 + x$ and $s_2 = 1 + x^2$ in $10s_1^4 s_2^3$ yields

$$10(1+x)^4(1+x^2)^3 = 10(1 + 4x + 6x^2 + 4x^3 + x^4)(1 + 3x^2 + 3x^4 + x^6)$$
$$= 10 + 40x + 90x^2 + 160x^3 + 220x^4 + 240x^5 + \cdots.$$

Similarly, from $15s_1^2 s_2^4$ we obtain

$$15(1+x)^2(1+x^2)^4 = 15(1 + 2x + x^2)(1 + 4x^2 + 6x^4 + 4x^6 + x^8)$$
$$= 15 + 30x + 75x^2 + 120x^3 + 150x^4 + 180x^5 + \cdots.$$

Substituting $s_1 = 1 + x$ and $s_3 = 1 + x^3$ in $20s_1 s_3^3$ gives

$$20(1+x)(1 + 3x^3 + 3x^6 + x^9) = 20 + 20x + 0x^2 + 60x^3 + 60x^4 + 0x^5 + \cdots.$$

The contribution from $20s_1 s_3 s_6$ is

$$20(1+x)(1+x^3)(1+x^6) = 20 + 20x + 0x^2 + 20x^3 + 20x^4 + 0x^5 + \cdots.$$

From $30s_2s_4^2$ we get

$$30(1+x^2)(1+2x^4+x^8) = 30+0x+30x^2+0x^3+60x^4+0x^5+\cdots.$$

and the final contribution, from $24s_5^2$, is

$$24(1+2x^5+x^{10}) = 24+0x+0x^2+0x^3+0x^4+48x^5+\cdots.$$

Summing up, we obtain

$$f_5(x) = \frac{1}{120}[120+120x+240x^2+480x^3+720x^4+720x^5+\cdots]$$
$$= 1+x+2x^2+4x^3+6x^4+6x^5+\cdots,$$

which, together with $g(5,m) = g(5, 10-m)$, yields Equation (79).

The $g(5,4) = 6$ nonisomorphic graphs with 5 vertices and 4 edges are illustrated in Fig. 10.5. While it is not difficult to see that no two of these six graphs are isomorphic, it may be less obvious that there can be no (lucky?) seventh. □

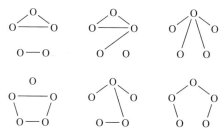

Figure 10.5

10.19 Corollary. *If $o(V) = n$, the total number of nonisomorphic graphs sharing vertex set V is*

$$f_n(1) = \frac{1}{n!}\sum_{p \in S_V} 2^{c(p)},$$

where $c(p)$ is the total number of cycles (including cycles of length 1) in the disjoint cycle factorization of $\tilde{p} \in S_V^{(2)}$.

Proof. If $o(V) = n$, the total number of nonisomorphic graphs sharing vertex set V is $f_n(1) = \sum_{m \geq 0} g(n,m)$. From Theorem 10.16, $f_n(x)$ is obtained by substituting $s_i = 1+x^i$, $1 \leq i \leq k$, in the cycle index polynomial

$$Z_n(s_1, s_2, \ldots, s_k) = \frac{1}{n!}\sum_{p \in S_V} s_1^{c_1(p)} s_2^{c_2(p)} \ldots s_k^{c_k(p)}.$$

Thus, $f_n(1)$ is obtained by substituting $s_i = 2$, $1 \le i \le k$, in the cycle index polynomial. It remains to observe that $c(p) = c_1(p) + c_2(p) + \cdots + c_k(p)$. ∎

10.20 Example. Suppose $n = 6$. Let $V = \{1, 2, 3, 4, 5, 6\}$. Listing the $C(6, 2) = 15$ elements of $V^{(2)}$ using hexadecimal numerals, we have

$\mathbf{1} = \{1, 2\}$, $\mathbf{2} = \{1, 3\}$, $\mathbf{3} = \{1, 4\}$, $\mathbf{4} = \{1, 5\}$, $\mathbf{5} = \{1, 6\}$,

$\mathbf{6} = \{2, 3\}$, $\mathbf{7} = \{2, 4\}$, $\mathbf{8} = \{2, 5\}$, $\mathbf{9} = \{2, 6\}$, $\mathbf{A} = \{3, 4\}$,

$\mathbf{B} = \{3, 5\}$, $\mathbf{C} = \{3, 6\}$, $\mathbf{D} = \{4, 5\}$, $\mathbf{E} = \{4, 6\}$, $\mathbf{F} = \{5, 6\}$.

If, $p = (12)(34)$, then $\tilde{p} = (\mathbf{27})(\mathbf{36})(\mathbf{48})(\mathbf{59})(\mathbf{BD})(\mathbf{CE})$. (Check it.) By Lemma 10.13, the 45 permutations of S_V having cycle structure $(ij)(xy)$ each contribute 1 to the coefficient of $s_1^3 s_2^6$ in the cycle index polynomial $Z_6 = Z_6(s_1, s_2, \ldots, s_{15})$. If $q = (12)(34)(56)$, then $\tilde{q} = (\mathbf{27})(\mathbf{36})(\mathbf{49})(\mathbf{58})(\mathbf{BE})(\mathbf{CD})$. So, the 15 permutations of cycle type $(ij)(xy)(st)$ each contribute 1 to the coefficient of $s_1^3 s_2^6$. These 15, together with the 45 we already have, yield a total coefficient of 60. Indeed,

$$Z_6(s_1, s_2, \ldots, s_{15}) = (s_1^{15} + 15 s_1^7 s_2^4 + 60 s_1^3 s_2^6 + 40 s_1^3 s_3^4 + 120 s_1 s_2 s_3^2 s_6$$
$$+ 40 s_3^5 + 180 s_1 s_2 s_4^3 + 144 s_5^3 + 120 s_3 s_6^2)/720. \quad (85)$$

Let's use Equation (85) to compute $g(6, 4)$, the number of nonisomorphic graphs with 6 vertices and 4 edges. From the binomial theorem, the contribution to the coefficient of x^4 in $(1 + x)^{15}$ is $C(15, 4) = 15 \times 7 \times 13$. The contribution from

$$15 s_1^7 s_2^4 = 15(1 + \cdots + 21 x^2 + \cdots + 35 x^4 + \cdots)(1 + 4 x^2 + 6 x^4 + \cdots)$$

is $15(6 + 84 + 35) = 15 \times 125$, for a subtotal (so far) of $15(91 + 125) = 15 \times 12 \times 18 = 12 \times 270$. From

$$60 s_1^3 s_2^6 = 60(1 + 3x + 3x^2 + x^3)(1 + 6x^2 + 15x^4 + \cdots),$$

we obtain $60(15 + 18) = 12 \times 5 \times 33$, for a (new) subtotal of $12 \times (270 + 165) = 12 \times 435$. The term

$$40 s_1^3 s_3^4 = 40(1 + 3x + \cdots)(1 + 4x^3 + \cdots)$$

contributes 12×40, bringing the subtotal to 12×475. From

$$120 s_1 s_2 s_3^2 s_6 = 120(1 + x)(1 + x^2)(1 + 2x^3 + x^6)(1 + x^6)$$

we get $240 = 12 \times 20$, raising the subtotal to 12×495. The term $40 s_3^5$ contributes nothing to the coefficient of x^4. The contribution from

$$180 s_1 s_2 s_4^3 = 180(1 + x)(1 + x^2)(1 + 3x^4 + \cdots)$$

Chap. 10 Edge Colorings

is $180 \times 3 = 12 \times 45$ bringing the subtotal to 12×540. Because neither $144s_5^3$ nor $120s_3s_6^2$ contribute to the coefficient of x^4, it remains to observe that $(12 \times 540)/720 = 9$. The nine nonisomorphic graphs with 6 vertices and 4 edges are illustrated in Fig. 10.6. ∎

Values of $g(n, m)$, for small n, appear in the table in Fig. 10.7.

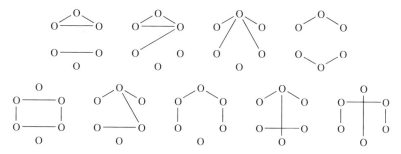

Figure 10.6. The nonisomorphic graphs with 6 vertices and 4 edges

						m					
n	0	1	2	3	4	5	6	7	8	9	10	11
1	1											
2	1	1										
3	1	1	1	1								
4	1	1	2	3	2	1	1					
5	1	1	2	4	6	6	6	4	2	1	1	
6	1	1	2	5	9	15	21	24	24	21	15	9
7	1	1	2	5	10	21	41	65	97	131	148	148
8	1	1	2	5	11	24	56	115	221	402	663	980

Figure 10.7. Numbers, $g(n, m)$, of graphs with n vertices and m edges

EXERCISES

1. Show that
   (a) $r(3, 10) \leq 46$. (b) $r(4, 6) \leq 43$.
   (c) $r(3, 10) \geq 19$. (d) $r(4, 6) \geq 16$.

2. Explain how the graph in Fig. 10.8 proves that $r(3, 4) > 8$.

3. Find the smallest integer value of $s \geq 2$ for which Erdös's Theorem gives a strictly better lower bound for $r(s, s)$ than Theorem 10.5.

4. Complete the proof of Theorem 10.7 by confirming that $2^{s/2}/s! < 1/2$ for all $s \geq 3$.

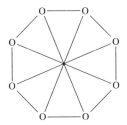

Figure 10.8

5. Let $s \geq 2$ be an integer. Suppose T is a fixed but arbitrary tree on $t \geq 2$ vertices. Let $N(s, T)$ be the smallest integer such that any graph G on $N(s, T)$ vertices contains an independent set of s vertices or a subgraph isomorphic to T. Prove that

   (a) $N(s, T) \geq (s - 1)(t - 1) + 1$.
   (b) $N(s, T) \leq (s - 1)(t - 1) + 1$.

6. Let T be a tree on k vertices. Let G be a graph with minimum vertex degree $\delta(G) \geq k - 1$. Prove that T is isomorphic to a subgraph of G.

7. Let $Z = \{1, 2, 3, 4, 5, 6\}$. Define $X = Z^{(2)}$ to be the set of all $C(6, 2) = 15$ 2-element subsets of Z and $Y = Z^{(3)}$ to be the set of all $C(6, 3) = 20$ 3-element subsets of Z. Let $V = X \cup Y$, $E = \{\{x, y\} : x \in X, y \in Y, \text{ and } x \subset y\}$, and $G = (V, E)$. Then G is a bipartite graph with bipartition $V(G) = V = X \cup Y$. Suppose each vertex of X is colored either black or white in some fixed but arbitrary way. Prove that there is a vertex $y \in Y$, each of whose neighbors is colored the same — in other words, such that each vertex adjacent to y is black or all of them are white.

8. Explain why

   (a) $g(n, 2) = 2, n \geq 4$. (b) $g(n, 3) = 5, n \geq 6$.
   (c) $g(n, m) = g(2m, m), n \geq 2m$.

9. Confirm that the expression for $\tilde{p}$ given in Fig. 10.3 is correct when

   (a) $p = (12)$ (b) $p = (12)(34)$ (c) $p = (1234)$

10. Let $p(x)$ be a polynomial of degree k. If $p(x) = x^k p(x^{-1})$ then $p(x)$ is a *reciprocal polynomial*.

    (a) Show that $f_4(x) = x^6 f_4(x^{-1})$.
    (b) Show that $f_5(x) = x^{10} f_5(x^{-1})$.
    (c) Show that $f_n(x)$ is a reciprocal polynomial, $n \geq 1$.

11. Compute

    (a) $Z_3(s_1, s_2, s_3)$.

(b) $f_3(x)$ using Theorem 10.16 and your answer to part (a).

(c) $f_3(x)$ using an illustration comparable to Fig. 10.2 for the nonisomorphic graphs on 3 vertices.

12 Consider the values of $g(n, m)$ tabulated in Fig. 10.7.

(a) Explain why the number of nonisomorphic graphs on 5 vertices can be obtained by summing the numbers row $n = 5$.

(b) Explain why the number of nonisomorphic graphs on 6 vertices cannot be obtained by summing the numbers in row $n = 6$.

13 Compute $\sum_{m=0}^{15} g(6, m)$, the number of graphs on 6 vertices

(a) using Fig. 10.7 and Equation (78).

(b) using (the proof of) Corollary 10.19 and Equation (85).

14 Confirm that $g(5, 6) = 5$ by computing the coefficient of x^6 in $f_5(x)$, in the manner of Example 10.20.

15 Compute the number of nonisomorphic graphs on 7 vertices.

16 Confirm, as in Example 10.20, that

(a) $g(6, 3) = 5$.   (b) $g(6, 5) = 15$.

17 Illustrate the six nonisomorphic graphs having 5 vertices and 6 edges.

18 Illustrate the six nonisomorphic graphs having 5 vertices and 5 edges.

19 Find the number of *connected* graphs on 5 vertices.

20 The *edge chromatic number*, $\chi'(G)$, is the minimum number of colors needed to color the edges of G so that adjacent edges are colored differently. (Equivalently, $\chi'(G)$ is the minimum number of matchings whose union is $E(G)$.)

(a) Prove that $\chi'(G) \geq \Delta(G)$.

(b) Show that $\chi'(G) = \Delta(G)$ for every connected graph on 4 vertices.

(c) If $G = K_3$, show that $\chi'(G) > \Delta(G)$.

(d) Find a connected graph $G \neq K_3$ for which $\chi'(G) > \Delta(G)$.[6]

21 Illustrate the 9 nonisomorphic graphs having 6 vertices and 11 edges.

22 Confirm, as in Example 10.20, that

(a) $g(6, 6) = 21$.   (b) $g(6, 7) = 24$.

23 Let $V = \{1, 2, 3, 4, 5, 6\}$. In the manner of Example 10.20, evaluate $\tilde{p}$ when

(a) $p = (1234)$.   (b) $p = (1234)(56)$.

---

[6] V. G. Vizing [On an estimate of the chromatic class of a p-graph, *Diskret. Analiz.* **3** (1964), 25–30], proved that $\chi'(G) \leq \Delta(G) + 1$ for all G.

24 Let $q(n, m)$ be the number of nonisomorphic multigraphs having n vertices and m edges and define the multigraph generating function

$$p_n(x) = \sum_{m \geq 0} q(n, m) x^m.$$

The variation of Theorem 10.16 for multigraphs is that $p_n(x)$ is obtained by substituting $s_i = (1 - x^i)^{-1}$, $1 \leq i \leq k = C(n, 2)$, in the cycle index polynomial $Z_n(s_1, s_2, \ldots, s_k)$; that is,

$$p_n(x) = Z_n(1/(1-x), 1/(1-x^2), \ldots, 1/(1-x^k))$$
$$= Z_n(1 + x + x^2 + x^3 + \ldots,$$
$$1 + x^2 + x^4 + x^6 + \ldots, \ldots, 1 + x^k + x^{2k} + x^{3k} + \ldots).$$

Use this result to obtain $p_4(x) = 1 + x + 3x^2 + 6x^3 + 11x^4 + 18x^5 + \ldots$.

25 Illustrate the nonisomorphic multigraphs having 4 vertices
   (a) and 2 edges. (b) and 3 edges.
   (c) and 4 edges. (d) and 5 edges.

26 Let f_1 and f_2 be orientations of the same graph G. Then the f_1-oriented graph G is isomorphic to the f_2-oriented graph G if (and only if) there exists an automorphism p of G such that $f_2(e) = f_1(\tilde{p}(e))$ for all $e \in E(G)$. Show that the number of nonisomorphic orientations of $G =$
   (a) C_3 is 2. (b) C_4 is 4. (c) C_5 is 4. (d) C_6 is 9. (e) K_4 is 4.

(*Hint*: Illustrate a system of distinct representatives for the nonisomorphic oriented graphs.)

Hints and Answers to Selected Odd-Numbered Exercises

CHAPTER 1

1. *Hint*: Number the vertices; then write down the 2-element subsets of $\{1, 2, 3, 4\}$ corresponding to edges.

3.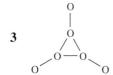

5. Six.

7. *Hint*: The "first theorem."

9. (a) *Hint*: The "first theorem." (b) *Hint*: $\Delta(G) \leq n - 1$.

11. *Hint*: The complement of the graph in part (a) is isomorphic to the graph in part (b).

13. *Hint*: Starting with a picture of K_n, illustrate G by coloring its edges dark. Then G^c is illustrated by the light edges.

15. *Hint*: The "first theorem."

17. *Hint*: Think of the rectangle with the long horizontal sides as the front face of a cube and the rectangle with the long vertical sides as its back face.

19. *Hint*: Consider their complements.

21. *Hint*: If $d(u) = \Delta(G) < n - 1$, then there exists some $w \in V(G)$ such that $uw \notin E(G)$. Let $P = [u, v, \ldots, w]$ be a shortest path in G from u to w. Show that if v is adjacent to every vertex in $U = \{x \in V(G): ux \in E(G)\}$, apart from itself, then $d(v) > d(u)$.

23. $G[1, 2, 3, 6]$, $G[1, 2, 4, 5]$, $G[1, 2, 5, 6]$, $G[1, 3, 4, 6]$, $G[2, 3, 4, 5]$, and $G[3, 4, 5, 6]$.

25. $\det(xI_n - A(G_1)) = x^4 - 4x^2 - 2x + 1 = \det(xI_n - A(G_2))$.

211

27 The common characteristic polynomial is $x^6 - 7x^4 - 4x^3 + 7x^2 + 4x - 1$.

29 Because $i = f^{-1}(j)$ if and only if $f(i) = j$, $\delta_{if^{-1}(j)} = \delta_{f(i)j} = \delta_{jf(i)}$. Thus, $P(f^{-1}) = P(f)^t$, and it suffices to prove part (b). To prove part (b), it suffices to show that $P(f)P(f)^t = I_n$ — in other words, that the (i, j)-entry of $P(f)P(f)^t$ is δ_{ij}. From the definition of matrix multiplication, the (i, j)-entry of $P(f)P(f)^t$ is

$$\sum_{k=1}^{n} \delta_{if(k)}\delta_{jf(k)} = \delta_{ij}$$

because f is one-to-one and onto.

31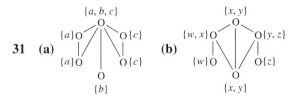

33 (e) *Hint*: From the first theorem of graph theory, $2m = k2^k$.

CHAPTER 2

1 (a,b) *Hint*: The Fundamental Counting Principle.

(c) $7 \times 5 \times 7 \times 5 = 1,225$.

3 (a) [graph] (b) $p(G, x) = x(x-1)(x-2)^3$.

5 *Hint*: Given that ω is an invariant, suppose $G \cong H$. Then $G^c \cong H^c$ and, hence, $\alpha(G) = \omega(G^c) = \omega(H^c) = \alpha(H)$, proving that α is an invariant.

7 *Hint*: Suppose f is an isomorphism from G to H. If $g: V(H) \to C$ is a proper r-coloring of H, then $g \circ f$ is a proper r-coloring of G. (Why?) Show that $g \to g \circ f$ gives a one-to-one correspondence between the proper r-colorings of H and the proper r-colorings of G.

9 (a) $p(P_3, x) = x(x-1)^2$. (b) $p(C_4, x) = x(x-1)(x^2 - 3x + 3)$.

(c) $p(C_5, x) = p(P_5, x) - p(C_4, x) = x(x-1)(x-2)(x^2 - 2x + 2)$.

11 $P_3 \oplus K_1$ and $2K_2$ share the chromatic polynomial $x^2(x-1)^2$.

13 *Hint*: $G = C_1^c \vee C_2^c \vee \cdots \vee C_r^c$.

Hints and Answers to Selected Odd-Numbered Exercises

15 *Hint*: G is an overlap of H_1 and H_2 in K_2, where $H_1 = K_4 - e$ is an overlap of K_3 with itself in K_2, and

$$H_2 = \begin{matrix} \circ - \circ - \circ \\ \diagdown \diagup \diagdown \diagup \\ \circ - \circ \end{matrix}$$

is an overlap of H_1 and K_3 in K_2. (A more direct approach is developed in Chapter 8.)

17 $p(W_5, x) = x(x-1)(x-2)(x-3)(x^2 - 4x + 5)$.

19 (a) $p(G, x) = x^{(5)} + x^{(4)} = x(x-1)(x-2)(x-3)^2$.
(b) $p(G, x) = x^{(6)} + 3x^{(5)} + 3x^{(4)} + x^{(3)}$
$= x(x-1)(x-2)(x^3 - 9x^2 + 29x - 32)$.

21 *Hint*: Theorem 2.13.

23 (a) $p(K_1 \vee K_3^c, x) = x^{(4)} + 3x^{(3)} + x^{(2)}$.

(b) $p(K_2^c \vee K_3^c, x) = x^{(5)} + 4x^{(4)} + 4x^{(3)} + x^{(2)}$.

(c) $p(K_3^c \vee K_3^c, x) = x^{(6)} + 6x^{(5)} + 11x^{(4)} + 6x^{(3)} + x^{(2)}$.

25 *Hint*: Using Exercise 9, show that $f(1) = 1$.

27 *Hint*: Suppose G is a graph. Show that $p(G \vee K_1, r) = rp(G, r-1)$. Use Exercise 26.

29 *Hint*: $f(x) = x(x-1)^2(x-3)^2(x-4)$.

31 *Hint*: Induction should be used only as a last resort. While it may help you decide what is true, it rarely sheds much light on why.

33 (a) *Hint*: How many of the nonisomorphic graphs on 4 vertices are trees?

(b) The nonisomorphic trees on 5 vertices have different degree sequences and different diameters.

(c) *Hint*: Among the trees on seven vertices, only P_7 has diameter 6, there are two trees of diameter 5, five of diameter 4, two of diameter 3, and one (namely, S_7) of diameter 2.

35 *Hint*: K_3 has a path of length 2.

37 (a) *Hint*: Exercise 36. (b) *Hint*: Part (a) and Exercise 36.

(d) *Hint*: Part (b) (e) *Hint*: Part (d)

39 *Hint*: If G is not a tree, it must have a cycle. Let e be an edge of the cycle. Show that $G - e$ is connected. Continue to remove edges until all cycles of G have been "broken." The resulting tree violates the condition established in Exercise 31.

41 *Hint*: Let $d(v_i, v_j) = D$. Use Chapter 1, Exercise 30, to show that the (i,j)-entry of $A(G)^k$ is 0, $1 \le k < D$, but the (i,j)-entry of $A(G)^D$ is nonzero. Deduce that $A(G)^D$ could not be a linear combination of $I_n, A(G), A(G)^2, \ldots, A(G)^{D-1}$. Conclude that the degree of the minimal polynomial of $A(G)$ is at least $D+1$. Finally, recall that the minimal polynomial of a real symmetric matrix A is the product of all terms of the form $(x - \lambda)$, as λ ranges over the distinct eigenvalues of A.

43 *Hint*: Let v be a pendant vertex of $T = (V, E)$ and $uv \in E$ the unique edge of T incident with it. Then $T - v = T[V \setminus \{v\}]$, the subgraph of T induced on the $k-1$ vertices different from v, is a tree. Show (by induction) that $T - v$ is isomorphic to a subgraph T' of G. Let u' be the vertex of T' corresponding to u. Show that, among the neighbors of u' in G, there is a vertex v' not belonging to $V(T')$.

45 (a) Find a counterexample.

(b) *Hint*: Recall the steps leading to the Balaban centric index.

47 (a) *Hint*: Let X_1 and X_2 be $1 \times k$ and $1 \times (n-k)$ matrices, respectively. Show that the $n \times 1$ matrix $X = (X_1, X_2)^t$, whose partitioning is consistent with Equation (15), is an eigenvector of $A(G)$ affording eigenvalue γ if and only if $Y = (X_1, -X_2)^t$ is an eigenvector of $A(G)$ affording $-\gamma$.

49 *Hint*: The parity of a binary codeword is odd or even according to whether the number of its 1's is odd or even.

51 *Hint*: G is the overlap of $G - u$ and K_{d+1} in K_d.

53 (b)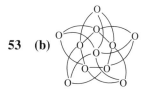

(d) *Hint*: Show that G_k is "triangle-free."

(e) *Hint*: Induction on k. Assume $\chi(G_k) = k$. Show that any proper k-coloring of G_k can be extended to a proper $(k+1)$-coloring of G_{k+1} by assigning to u_i the same color as v_i, $1 \le i \le n$, and giving w a new color. Because G_{k+1} has an induced subgraph isomorphic to G_k, $k \le \chi(G_{k+1}) \le k+1$. If $\chi(G_{k+1}) = k$, we may assume color c_k is assigned to vertex w. Because $wu_i \in E(G_{k+1})$, $1 \le i \le n$, no u_i is colored c_k. Recolor the v's, if necessary, so that v_i has the same color as u_i, $1 \le i \le n$. This produces a proper $(k-1)$-coloring of G_k.

CHAPTER 3

1 (a) G is disconnected; $\varepsilon(2K_3) = 0$. (b) $\varepsilon(G) = 2$.

(c) $\varepsilon(C_6) = 2$. (d) $\varepsilon(K_{3,3}) = 3$.

Hints and Answers to Selected Odd-Numbered Exercises

3 Many examples are possible, e.g.,

(a) $K_4 \oplus K_4$.

(b) Add an edge to $K_4 \oplus K_4$.

(c) Let $H = K_4 \oplus K_4$. Let G be the connected graph obtained from H by coalescing a vertex from one component with a vertex of the other.

(d) Let G be an overlap of K_4 and K_4 in K_2.

5 *Hint*: Show that u and v lie on a cycle of G if and only if there are two "internally disjoint" paths in G from u to v.

7 (a) *Hint*: Let f be an isomorphism from G to H. Show that S is a separating set of G if and only if $f(S) = \{f(v): v \in S\}$ is a separating set of H.

9 *Hint*: Theorem 3.13.

11 Let $G =$ [graph] and $P = [u, x, y, v]$.

13 (b) *Hint*: Exercise 10.

15 *Hint*: Corollary 3.11.

17 False. Take $G = C_n$. This explains why Theorem 3.15 is "if and only if" while Theorem 3.20 is not.

19 [graph]

21 *Hint*: A tree is a connected acyclic graph.

23 *Hint*: "Subdivide" the two edges as in Exercise 16.

25 $\kappa(G) = 2 = \psi_G(u, w); \psi_G(u, v) = 3$.

27 False. Let $G =$ [graph].

29 Suppose $S \subset V(G)$, $o(S) < k$, and $G - S$ is disconnected. Let C be a smallest component of $G - S$. If u is a vertex of C, show that $d_G(u) < \delta(G)$.

31 $p(G, x) = x(x - 1)^4(x - 2)^3(x^2 - 3x + 3)$.

33 *Hint*: Do part (c) first.

35 The 3-connected graphs on 5 vertices are K_5, $K_5 - e$, and $K_5 - M$, where M is a set consisting of two nonadjacent edges of K_5. The seven other blocks on 5 vertices all satisfy $\delta(G) = 2$.

37 (b) The girth of the Petersen graph is 5.

(c) The circumference is 9. It is not hard to find a cycle of length 9. With respect to the numbering of the vertices in Fig. 3.8, $C = \langle 1, 2, 3, 4, 5, 6, 7, 8, 9 \rangle$ is such a cycle. (Indeed, to answer one of the questions posed at the end of the Chapter, there exists a cycle of G containing *any* prescribed set of 9 vertices.) It is a much harder problem to prove that the Petersen graph does not contain a cycle of length 10. This is the Hamiltonian cycle problem, the subject of Chapter 5.

39 *Hint*: Let $U = \{u, v\}$. If $e = uv \in E(G)$, show that $H = G - e$ is connected. If $[u, z_1, \ldots, z_r, v]$ is a path in H from u to v, then $\langle u, z_1, \ldots, z_r, v \rangle$ is a cycle of G containing u and v. If $uv \notin E(G)$ use Menger's Theorem to obtain two internally disjoint paths in G from u to v.

41 (a), (d), and (figures)

(e) $p(H, x) = x^4 - 4x^3 + 5x^2 - 2x$.

43 (b) Show that $f(1) = -1$ and $f(2) = +1$, where $f(x) = x^3 - 5x^2 + 10x - 7$.

CHAPTER 4

1 *Hint*: Lemma 2.29.

3 *Hint*: Show that $\delta(G) \geq 3$ for any triangulated plane graph on $n \geq 7$ vertices.

5 (a), (b) (figures)

7 $K_{3,3}$ is bipartite.

9 (a) *Hint*: From geometrical considerations, each vertex of a fullerene must be at the intersection of exactly 3 faces. Thus, the number of vertices of a fullerene with p pentagonal and h hexagonal faces is $V = (5p + 6h)/3$. Find a similar formula for the number of edges. (If G is a connected, *cubic*,[1] plane graph, and r_k is the number of regions of G bounded by a cycle of length k, then $3r_3 + 2r_4 + r_5 = 12 + r_7 + 2r_8 + 3r_9 + \cdots + (k-6)r_k + \cdots$.)

(c) $h = (n - 20)/2$.   (d) $h = 20$.

11 (a) Use Kuratowski's Theorem.

(c) *Hint*: Corollary 4.3.

15 C_4 is bipartite.

---

[1] "Cubic" is a synonym for "3-regular," referring to graphs each of whose vertices has degree 3.

Hints and Answers to Selected Odd-Numbered Exercises

17 (a) 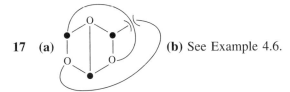 (b) See Example 4.6.

19 *Hint*: Many proofs are possible, e.g., let $u, v \in V(G_d)$, where u is the vertex corresponding to the unbounded region of G. Show that there is a path in G_d from u to v.

21 The Heawood graph: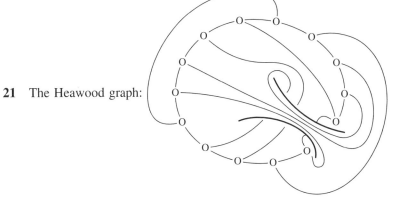

23 *Hint*: By Theorem 4.21, $\gamma(K_{12}) = 6$ and $\gamma(K_{13}) = 8$.

25 $r(K_7) = 14$.

27 (c) $\gamma(K_{7,7}) = 7$ (coincidentally).

29 *Hint*: The "outside" region of the toroidal embedding of K_5 in Example 4.17 has a perimeter of length 8.

31 (a) The adjacency eigenvalues of $G = P_3$ are $\sqrt{2}, 0, -\sqrt{2}$. Thus, $H(G) = 2\sqrt{2} \doteq 2.83$.

(b) Numbering the vertices of $G = C_6$ consecutively around the cycle, show that $X = (1, 1, 1, 1, 1, 1)^t$, $Y = (1, 0, -1, -1, 0, 1)^t$, and $Z = (1, 1, 0, -1, -1, 0)^t$ are (linearly independent) eigenvectors of $A(G)$ affording eigenvalues 2, 1, and 1, respectively. Use Chapter 2, Exercise 47, to show that, multiplicities included, the eigenvalues of $A(G)$ are 2, 1, 1, -1, -1, and -2. Thus $H(G) = 8$.

33 $H = o—o—o{\Big\langle}\begin{smallmatrix}o\\o\end{smallmatrix}$

35 (a) A regular octahedron has 8 equilateral triangular faces and 6 vertices, each formed by the intersection of 4 faces.

(d) *Hints*: (1) Identify a "spanning cycle" of G; and (2) show that G satisfies $m = 3n - 6$.

(f) See the hints to part (d).

(g) *Hint*: Show that $\omega(H) = 4$.

(h) *Hint*: $\chi(G) \neq \chi(H)$.

(i) *Hint*: Part (b) and Chapter 2, Exercise 19(b).

(j) *Hint*: Begin by showing that H is the overlap of $K_5 - e$ with itself in K_4.

CHAPTER 5

1 **(a)** Because the edge consisting of the bottom two vertices is a bridge, $\kappa(G) = 1$.

(b) If S is the set consisting of the top three vertices then $\xi(G - S) = 4$.

(c) Yes.

5 *Hint*: Apply Corollary 5.10 to $G = H \vee K_1$.

7 *Hint*: Exercise 6.

9 *Hint*: Corollary 5.9.

11 **(a)** 29. **(b)** 25. **(c)** 24. **(d)** 26.

13

15 24.

17 The Petersen graph.

19 *Hint*: Show that for any two vertices u and v of G, $H = G - u - v$ has a connected spanning subgraph.

21 *Hint*: $s_3 = 6$ and $s_i = 0$, $i \neq 6$; $t_4 = 3$ and $t_i = 0$, $i \neq 4$. (Don't forget the unbounded region.)

23 Subdivide an edge of $K_{3,3}$.

27 **(a)** *Hint*: Theorem 5.3.

31 See Chapter 1, Exercise 22 for the definition of "multigraph," and see Definition 4.7 for that of "subdivide."

Hints and Answers to Selected Odd-Numbered Exercises

CHAPTER 6

1 (a) 3. (b) 5.

3 There are 4 Kekulé structures in anthracene and 5 in phenanthrene, so phenanthrene should be the most stable.

5 (a) *Hint*: Example 6.8. (c) *Hint*: Example 6.11.

7 (a) *Hint*: Example 6.8. First use the quadratic formula to find the squares of the roots.

(b) *Hint*: Example 6.11. Use the quadratic formula to find the squares of the roots, then use a calculator. (Two decimal place accuracy should suffice.)

9 (a) $M(K_5 - e, x) = x^5 - 9x^3 + 12x$.

11 *Hint*: Theorem 6.12.

13 *Hint*: Let $X = \{x_1, x_2, \ldots, x_m\}$ and $Y = \{y_1, y_2, \ldots, y_n\}$. Let $G = (V, E)$ be the bipartite graph with color classes X and Y such that $x_i y_j \in E$ if and only if $a_{ij} = 1$. Show that the theorem on term ranks is equivalent to $\beta(G) = \mu(G)$.

15 *Hint*: The symmetric difference of sets A and B is $A \triangle B = (A \cup B) \setminus (A \cap B)$. Suppose G contains an M-augmenting path P. Denote by $E(P)$ the set of edges along P and show that $M^* = E(P) \triangle M$ is a matching of G satisfying $o(M^*) > o(M)$. Conversely, suppose M and M^* are matchings of G with $o(M^*) > o(M)$. Show that $M \triangle M^*$ is comprised of the edges of a family of M-alternating paths and cycles. Because the cycles contain an equal number of edges from M and M^*, at least one of the paths must be M-augmenting.

19 (b) $x(x-1)^6(x-2)^2$ (c) $x^9 - 10x^7 + 29x^5 - 25x^3 + 5x$

21 (b) *Hint*: Γ_4 has 11 vertices and 14 edges.

(c) *Hint*: If G and $H \in V_n$ have the same number of edges, they cannot be adjacent in Γ_n. (Why?)

23 (a) *Hint*: per($A(G)$) counts the permutations p for which the *diagonal product* $\prod a_{t p(t)} \neq 0$. Consider the disjoint cycle factorizations of such permutations.

(b) *Hint*: Because every vertex of G is covered by a perfect matching, every ordered *pair* of perfect matchings determines a unique *bipartite* Sachs subgraph H of G. Conversely, given any bipartite Sachs subgraph H of G, there are $2^{c(H)}$ ordered pairs of perfect matchings to which it corresponds; that is, the number of ordered pairs of perfect matchings of G is

$$K(G)^2 = \sum_{H \in S(G)} 2^{c(H)}.$$

25 *Hint*: Let $S = \{u, v, w\}$ and $T = X \backslash N_G(S)$. Observe that $N_G(T) = Y \backslash S$.

27 *Hint*: Suppose $G = (V, E)$ has a matching N that covers a set T of vertices of G, where $o(T) > o(S)$. Let $H = (V, M \cup N)$. Show that one of the components of H is a path P, neither of whose pendant vertices is covered by M. Conclude that the symmetric difference (see the hint to Exercise 15) of M and $E(P)$, the set of edges of P, is a matching of G that covers S *and* two vertices of T not contained in S.

29 **(a)** *Hint*: Menger's Theorem.

(b) Let $G = K_5 - e$.

31 *Hint*: From matrix theory, c_k is $(-1)^k$ times the sum of the determinants of all $k \times k$ principal submatrices of $A(G)$. A $k \times k$ principal submatrix of $A(G)$ is the adjacency matrix $A(H)$ of the subgraph H of G induced on the corresponding k vertices. If $A(H) = (a_{ij})$, then

$$\det(A(H)) = \sum_{p \in S_k} \varepsilon(p) \prod_{t=1}^{k} a_{t p(t)},$$

where S_k denotes, not the star on k vertices, but the set of all $k!$ permutations on $\{1, 2, \ldots, k\}$, and $\varepsilon(p)$ is not the edge connectivity of a graph but the *signum* of permutation p; that is, $\varepsilon(p) = k - s(p)$, where $s(p)$ is the number of cycles in the disjoint cycle factorization of p.

33 **(a)** $x^4 - 3x^2 + 1$.

(b) $x^4 - 4x^2$. (Compare with $M(C_4, x) = x^4 - 4x^2 + 2$.)

(c) $x^4 - 6x^2 - 8x - 3$. (Compare with $M(K_4, x) = x^4 - 6x^2 + 3$.)

35 **(b)** *Hint*: Divide the k-matchings of $A_n \cong (A_{n-2} \oplus p) \vee w$ into three types.

Type 1: If w is not covered by an edge of the matching, then all k edges come from A_{n-2}.

Assuming $e = vw$ is an edge of the matching, it may be:

Type 2: that $v = p$. If so, the remaining $k - 1$ edges of the matching come from $A_n - p - w \cong A_{n-2}$.

Type 3: that $v \neq p$. If so, the remaining edges come from $(A_n - w) - v \cong (A_{n-2} \oplus p) - v$ and, hence, from $A_{n-2} - v$. Use Exercise 16, to show that there are $(n - 2k) \times a(n - 2, k - 1)$ matchings of this type.

37 **(b)** $K_{4,4}$

Hints and Answers to Selected Odd-Numbered Exercises

39 (a) [two graph diagrams with vertices labeled u, v]

CHAPTER 7

1. (5), (4, 1), (3, 2), (3, 1, 1), (2, 2, 1), (2, 1, 1, 1), and (1, 1, 1, 1, 1). [*Remark*: When exhibiting partitions, it is common to use superscripts to denote multiplicities, writing, for example, $(2, 1^3)$ in place of (2, 1, 1, 1).]

3. *Hint*: Figure 7.9.

5. **(a)** $r(\pi) = 3$. **(b)** $r(\pi) = 2$. **(c)** $r(\pi) = 4$. **(d)** $r(\pi) = 3$.

7. $(6, 1^5) = (6, 1, 1, 1, 1, 1)$ and $(4, 3^2, 1) = (4, 3, 3, 1)$.

9. **(b)** *Hint*: Use vertices $p, v, w,$ and q. [graph diagram with vertices p, u, v, w, q]

11. *Hint*: Figure 1.9.

13. **(a)** The residue is 4.

    (c) Among the possibilities are (7, 5, 4, 3, 3, 2, 1, 1) and (6, 5, 4, 4, 4, 2, 1), but there are other correct answers.

15. None, 7 is odd.

17. *Hint*: Figure 7.13.

19. **(a)** 2. **(b)** 9. **(c)** 11.

21. *Hint*: $d(G_1) = (4, 4, 2, 2, 2)$ and $d(G_2) = (4, 3, 3, 3, 1)$.

23. **(a)** [two graph diagrams]

23. **(b)** [three graph diagrams]

23 (c)

25 (a) *Hint*: Suppose $\pi = (\pi_1, \pi_2, \ldots, \pi_n)$, where $\pi_n > 0$. Split the argument into two cases according to whether $\pi_n = 1$ or $\pi_n > 1$.

(b) *Hint*: Use part (a).

27 *Hint*: Show that moving boxes down in $F(\gamma)$ corresponds to moving boxes up in its transpose, $F(\gamma^*)$.

29 *Hint*: Many approaches are possible. A proof relying on degree sequences goes as follows. Suppose G has p vertices and q edges. Let $\mu = d(G) \vdash 2q$ and $\nu = d(K_n \vee G) \vdash 2m$. Show that the first n rows and columns of $F(\nu)$ contain $p + n - 1$ and $p + n$ boxes, respectively. Show that $F(\mu)$ is obtained from $F(\nu)$ by removing its first n rows and columns.

31 *Hint*: Suppose $d(G) = (5, 2, 2, 1, 1, 1)$ and $d(H) = (3, 3, 2, 2)$. Show that $d(G \vee H)$ is not a threshold partition.

33 *Hint*: The center of a connected threshold graph consists of its dominating vertices.

35 *Hint*: The Threshold Algorithm affords one way to approach this exercise.

37 *Hint*: Mimic the proof of Theorem 7.22.

39 *Hint*: Consider G^c.

CHAPTER 8

1 (a) $\beta \vdash 10$ does not weakly majorize $\alpha \vdash 11$;

(b) $\beta \succeq_w \alpha$ (c) $\beta \succeq_w \alpha$ (f) $\beta \succ \alpha$

3 (a) $\alpha(\pi) = (5, 3, 1) = \beta(\pi)$

(b) $\alpha(\pi) = (5, 3); \beta(\pi) = (6, 4)$

(c) $\alpha(\pi) = (4, 1); \beta(\pi) = (6, 4)$

(d) $\alpha(\pi) = (5, 3); \beta(\pi) = (6, 2)$

5 (a) (a), (b), and (d) (b) (a) and (d) (c) (a)

7 $(5, 3, 2, 2, 1, 1), (4, 4, 2, 2, 2)$, and $(4, 3, 3, 3, 1)$.

9 (a) 5 (b) 6 (c) 8 (d) 10

11 *Hint*: π is graphic if and only if $\beta(\pi) \succeq_w \alpha(\pi)$, and split if and only if $\beta(\pi) \succ \alpha(\pi)$.

13 (graph figure)

15 *Hint*: The Threshold Algorithm.

17 Because every subgraph (induced or not) of a bipartite graph is bipartite, it suffices to prove that $\chi(G) = \omega(G)$ for every bipartite graph G. So, let G be a bipartite graph. Then, by definition, $\chi(G) \leq 2$. Because G has no odd cycles, $K_3 = C_3$ cannot be a subgraph of G. Hence $\omega(G) \leq 2$. Because $\chi(G) \leq \omega(G)$, it remains to prove that $\chi(G) = 1$ implies $\omega(G) = 1$.

19 *Hint*: Theorem 8.25.

21 *Hint*: Theorem 8.9.

23 *Hint*: Corollary 8.26.

25 *Hint*: Figure 8.14.

27 *Hint*: $\{v\}$ is both a clique and an independent set, $v \in V$.

29 P_5 is not a split graph.

31 *Hint*: Show that

$$\sum_{i=1}^{k} \pi_i = \sum_{i=1}^{k} \alpha_i + k(k-1)/2 \quad \text{and} \quad \sum_{i=k+1}^{n} \min\{k, \pi_i\}$$

$$= \sum_{i=1}^{k} \beta_i - k(k-1)/2,$$

$1 \leq k \leq f(\pi)$, where $\alpha = \alpha(\pi)$ and $\beta \equiv \beta(\pi)$.

33 **(a)** *Hint*: Chapter 7, Exercise 26.

(b) *Hint*: The proof of Theorem 8.9.

35 *Hint*: Exercise 34.

37 **(a)** $p(G, x) = x(x-1)(x-2)^5$.

(b) $p(G, x) = x(x-1)(x-2)^4(x-3)$.

39 *Hint*: Chapter 7, Exercise 22, and Corollary 8.28. (Compare with Chapter 2, Exercise 38.)

CHAPTER 9

1 Compute or recall that $p(G, x) = x(x-1)^2(x-2)$. Show that $p(G, -1) = 12$. Thus, of the $2^4 = 16$ orientations of G, 12 are acyclic. Explicitly diagram either the 12 acyclic orientations, or the four orientations that afford a directed cycle.

3 *Hint*: $QQ^t = L(G)$.

5 (a) $s(C_4) = (4, 2, 2, 0)$ (b) $s(C_4 + e) = (4, 4, 2, 0)$ (c) $s(S_4) = (4, 1, 1, 0)$ (d) $s(S_4 + e) = (4, 3, 1, 0)$

7 (a) $a(C_4) = 2 = \kappa(C_4)$ (d) $a(P_4) = 2 - \sqrt{2} < 1 = \kappa(P_4)$

9 *Hint*: Use Equation (66).

11 (a) *Hint*: $L(K_n^c) = 0$. (b) *Hint*: Equation (65).

13 (a) $s(G) = (6, 4, 4, 3, 1, 0)$ (b) $s(G^c) = (5, 3, 2, 2, 0, 0)$ (d) $s(G \vee G^c) = (12, 12, 11, 10, 10, 9, 9, 8, 8, 7, 6, 0)$ (e) $s(H) = (6, 5, 4, 3, 2, 0)$ (f) $s(H^c) = (4, 3, 2, 1, 0, 0)$

15 (a) *Hint*: $K_{r,s} = K_r^c \vee K_s^c$. (b) *Hint*: Equation (65).

17 (a) *Hint*: Lemma 7.27. (b) *Hint*: Definition 5.16.

19 *Hint*: Exercise 18 or the fact that $H_1^c = H_2$.

21 *Hint*: $\sum_{i=1}^n \lambda_i(G) = \text{trace } L(G) = \sum_{v \in V(G)} d_G(v)$.

23 *Hint*: Exercise 22.

25 *Hint*: Find an eigenvector.

27 (b) *Hint*: Equation (65).

29 *Hint*: Chapter 1, Exercise 20; and Chapter 7, Exercise 22.

33 *Hint*: Consider their complements.

35 (a) *Hint*: Use Exercise 32 and the fact that rank Q = rank QQ^t.

(b) *Hint*: The dimension of W is the "nullity" of Q. By "Sylvester's Law of Nullity," rank Q + nullity Q = number of columns of Q.

(d) *Hint*: Part (c) (e) *Hint*: Part (c).

(g) *Hint*: Part (f).

37 $K_3 \oplus K_2$ is a 4-edged subgraph of K_5.

39 *Hint*: Consider a proper coloring $f: V(G) \to C$, where C is the set of "colors" $\{1, 2, \ldots, k\}$. Orient each edge so that the vertex having the larger color is its head.

41 *Hint*: Prove that every *forest* on n vertices is isomorphic to a spanning subgraph of A_n. The only forests on $n \leq 2$ vertices are $K_1 = A_1$, $P_2 = A_2$, and K_2^c. Suppose F is a forest on $n \geq 3$ vertices. If $F = K_n^c$, then F is a spanning subgraph of A_n. Otherwise, let u be a vertex of F of degree 1 and w its neighbor. Let $S = \{u, w\}$. By induction, $F - S$ is isomorphic to

a spanning subgraph of A_{n-2}. Use Chapter 2, Exercise 37 to complete the proof.

CHAPTER 10

1. *Hint*: Mimic Example 10.6.
3. $s = 16$.
5. (a) *Hint*: See the proof of Theorem 10.5.
   (b) *Hint*: Induction on $s + t$.
7. *Hint*: Show that this is another model of the cocktail party observation.
11. (a) $Z_3(s_1, s_2, s_3) = (s_1^3 + 3s_1 s_2 + 2s_3)/6$.
    (b) $f_3(x) = 1 + x + x^2 + x^3$.
13. *Answer*: 156.
15. *Hint*: Use Fig. 10.7. *Answer*: 1044. (The number of nonisomorphic graphs on 10 vertices exceeds 12 billion.)
17. *Hint*: They are the complements of the graphs in Fig. 10.5.
19. *Answer*: 21.
21. *Hint*: They are the complements of the graphs with 6 vertices and 4 edges.
23. (a) $\tilde{p} = $ (16A3) (27) (48BD) (59CE)
    (b) $\tilde{p} = $ (16A3) (27) (49BE) (58CD)
25. (a) *Hint*: From Exercise 24, there are $q(4, 2) = 3$ such multigraphs (two of which are graphs).

    (b)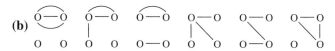

    (c) *Hint*: From Exercise 24, there are $q(4, 4) = 11$ such multigraphs (two of which are graphs).

    (d) *Hint*: From Exercise 24, there are $q(4, 5) = 18$ such multigraphs (one of which is a graph).

Bibliography

GENERAL

J. A. Bondy and U. S. R. Murty, *Graph Theory with Applications*, North-Holland, Amsterdam, 1976.

G. Chartrand and L. Lesniak, *Graphs & Digraphs,* 2nd ed., Wadsworth & Brooks/Cole, Pacific Grove, CA, 1986.

F. Harary, *Graph Theory*, Addison-Wesley, Reading, MA, 1972.

N. Hartsfield and G. Ringel, *Pearls in Graph Theory*, Academic Press, New York, 1990.

D. B. West, *Introduction to Graph Theory,* Prentice-Hall, Englewood Cliffs, NJ, 1996.

SPECIAL TOPICS

D. Bonchev and D. H. Rouvray, eds., *Chemical Graph Theory: Introduction and Fundamentals*, Abacus Press/Gordon & Breach, New York, 1991.

R. C. Brigham and R. D. Dutton, A compilation of relations between graph invariants, *Networks* **15** (1995), 73–107; Supplement I, *Ibid.* **21** (1991), 421–455.

G. L. Chia, A bibliography on chromatic polynomials, *Discrete Math.* **172** (1997), 175–191.

D. M. Cvetković, M. Doob, and H. Sachs, *Spectra of Graphs*, 3rd ed., Johann Ambrosius Barth Verlag, Heidelberg, 1995.

M. C. Golumbic, *Algorithmic Graph Theory and Perfect Graphs*, Academic Press, New York, 1980.

R. J. Gould, Updating the hamiltonian problem — a survey, *J. Graph Theory* **15** (1991), 121–157.

R. L. Graham, M. Grötschel, and L. Lovász, *Handbook of Combinatorics*, MIT Press/North-Holland, Cambridge, MA, 1995.

R. L. Graham and J. H. Spencer, Ramsey theory, *Scientific American*, July 1990, pp. 112–117.

I. Gutman and O. E. Polansky, *Mathematical Concepts in Organic Chemistry*, Springer-Verlag, Berlin, 1986.

F. Harary and E. M. Palmer, *Graphical Enumeration*, Academic Press, New York, 1973.

T. R. Jensen and B. Toft, *Graph Coloring Problems*, Wiley, New York, 1995.

L. Lovász and M. D. Plummer, *Matching Theory*, Annals of Discrete Mathematics 29, North-Holland, Amsterdam, 1986.

N. V. R. Mahadev and U. N. Peled, *Threshold Graphs and Related Topics*, Annals of Discrete Mathematics 56, Elsevier, Amsterdam, 1995.

G. Pólya and R. C. Read, *Combinatorial Enumeration of Groups, Graphs, and Chemical Compounds*, Springer, New York, 1987.

N. Trinajstić, *Chemical Graph Theory* (in two volumes), CRC Press, Boca Raton, FL, 1983.

Index

Acyclic
  orientation   172
  polynomial   106
Adjacency matrix   11
Adjacent   3
Adjugate   177
Algebraic connectivity   180
Algorithm   34, 86, 139ff
Alkane   33
Alternant hydrocarbon   29
Altman, M.   187
Anderson, I.   115
Antiregular graph   17, 40, 80, 96, 123, 143, 169, 191, 193
Appel, K.   69
Archimedes   77
Automorphism   5

Bagel graph   72
Balaban, A. T.   34
Balaban centric index   35, 167
Balaban's Algorithm   34
Balaji, K.   166
Barnard, S. T.   180
Beard, J. T. B., Jr.   144
Benzene   80, 103ff
Berge, C.   62, 159, 168
Biggs, N.   192, 193
Binary codeword   19
Bipartite graph   29ff, 42, 66, 112ff
Bipartition   30
Birkhoff, G. D.   82
Block   48
  of a graph   48
  of a sample space   114
Bonchev, D.   33, 227
Bondy, J. A.   227
Boundary   66, 74
Bridge   46
Brigham, R. C.   227
Broken
  cycle   57
  wheel   37
Brooks, R. L.   24
Brooks's Theorem   24
Buckminsterfullerene   117

Cameron, P. J.   62
Carbon skeleton   33ff, 103ff
Cauchy–Binet Theorem   178
Cayley, A.   69
Center   41, 144
Chartrand, G.   93, 227
Chelnokov, V. M.   194
Chemical graph   80
Chia, G. L.   227
Chord   66, 92, 156
Chordal graph   156
Chromatic
  number   23
  polynomial   29, 57, 82, 107, 162ff, 172
  reduction   27, 107
Chromatically equivalent   26, 33, 36, 163
Chvátal graph   89, 96, 190
Chvátal, V.   88, 89, 135, 165
Chvátal's Theorem   90
Circumference   60
Clar postulate   117
Classical adjoint   177
Clique   22
  number   22
Closed walk   73, 101
Closure of G 86
Coalescence   38
Cocktail party graph   80, 130
Codeword   19
Color class   22
Coloring   21
Comparability graph   189
Complement   9

Complete
  bipartite graph   30
  graph   9
  list of invariants   7
Component   8
Congruent matrices   177
Conjugate partition   133
Connected
  component   8
  graph   8
Connectivity   47
  edge   45
Contractible   76
Contraction   27, 76, 172
Cook, S. A.   7
Coplanar   77
Cotree   40
Coulson, C. A.   42
Cover(ed)
  edges   111
  vertices   105, 111
Covering   111
  minimal   111
  minimum   111
  number   112
Critically k-chromatic   42, 168
Cross(ing edges)   63
Cube graph   19, 42, 61, 80, 98, 123
Cubic graph   216
Curl, R.   76
Cut−edge   46
Cut−vertex   47
Cvetković, D. M.   42, 77, 122, 227
Cycle
  even   24
  graph   2
  in G 30
  index polynomial   202
  odd   24
  space   192

Decomposable graph   183ff
Deficiency of a matching   121, 123
Degree   4
Diameter   31
Direct sum of matrices   13
Directed
  arc   171
  cycle   172
  graph   171
  path   172
Disconnected graph   8
Disconnecting set of edges   45
Disjoint cycle factorization   110ff, 205

Distance
  between codewords   19
  between graph vertices   30, 187
Dominating
  degree sequence   90
  vertex   4
Doob, M.   227
Dorris, Ann D.   144
Dual pseudograph   69−70, 94ff
Durfee square   133
Dutton, R. D.   227

Eccentricity   41, 144
Edelman, P. H.   164
Edge   3
  chromatic number   209
  connectivity   45
  covering   121
  covering number   121
  crossing   63
  cut   60
  subgraph   26
Egerváry, E.   112
Egerváry−König Theorem   112
Embedding   63
Energy   80
Equivalent
  cycles   84
  graphs   135
  partitions   126
Erdös, P.   167, 198
Erdös's Theorem   198
Euclid   69, 130
Euler, L.   64, 100, 150
Euler tour   99−101
Eulerian graph   101
Euler's Formula
  for plane graphs   64
  for connected graphs   64, 65, 72
Expander graph   113
Expansive set of vertices   113

Falling factorial   22
Fáry, I.   76
Favaron, O.   127
Favorinus   46
Ferrers diagram   131
Ferrers, N. M.   131
Fibonacci number   118
Fiedler, M.   180
First theorem of graph theory   7
Five−Color Theorem   67
Flow   192
Foldes, S.   158

Index

Forbidden (sub)graph 66, 76, 89, 130, 138, 156, 159, 184
Forest 31
Four–Color Theorem 69
Fowler, P. W. 80
Franklin, B. 64
Fuller, R. B. 77
Fullerene 76, 80
Fundamental Counting Principle 21

Gallai, T. 167, 193
Garey, M. R. 81
Gasharov, V. 144
Generating function 19, 150, 199
Genus
 of a graph 72
 of a surface 71
Girth 60
Godsil, C. 120, 186
Golumbic, M. C. 144, 227
Gould, R. J. 227
Graham, R. 227
Graph 3
 generating function 199
 isomorphism problem 7
Graphic partition 126ff, 149
Griggs, J. R. 128
Grinberg, E. J. 93
Grinberg's Theorem 93
Grone, R. 164, 190
Grötschel, M. 227
Grötzsch, H. 81
Grund, R. 15
Guthrie, Francis 69, 92
Guthrie, Frederick 69
Gutman, I. 133, 149, 188, 227
Gyárfás, A. 193

Haken, W. 69
Hakimi, S. L. 126
Hall, P. 114
Hall's Theorem 113–114
Hamilton, W. R. 69, 83
Hamiltonian
 connected 97
 cycle 84
 graph 84
 path 84
Hammer, P. L. 135, 158, 165
Hamming distance 19
Harary, F. 143, 227
Hartsfield, N. 227
Hässelbarth, W. 133
Havel, V. 126

Havel–Hakimi
 residue 127
 Theorem 126
Head (of an oriented edge) 171
Heath, J. R. 76
Heawood graph 78, 217
Heawood, P. 69, 75
Heawood's Theorem 75
Heilmann, O. J. 106
Henderson, P. B. 136, 144
Hendrickson, B. 180
Hermite, C. 110
Hermite polynomials 110
Holton, D. A. 186
Homeomorphic graphs 67
Hosoya, H. 33, 106
Hosoya topological index 118
Hückel
 graph 80
 molecular orbital method 80

Ibaraki, T. 158
Incident 3
Independence number 22
Independent
 list of invariants 7
 set of vertices 22
Induced
 permutation 201
 subgraph 8
Infinite graph 3
Internally disjoint paths 52
Intersection
 graph 19
 number 19, 41
Interval graph 143, 166
Invariant 5
Isolated vertex 9
Isomer 33, 104
Isomorphic 5
Isomorphism 5
Isospectral graphs 183, 186, 192

Jackson, B. 62
Jensen, T. R. 81, 227
Johnson, C. R. 164
Johnson, D. S. 81
Join
 of graphs 23
 product 39
Jordan Curve Theorem 92
Jovanović, A. 77

Kant, I. 100
Karp, R. M. 7

k-connected   50
Kekulé structure   106
Kekulé von Stradonitz, F. A.   104ff
Kel'mans, A. K.   194
Kempe, A.   69
Kierstead, H. A.   43
Kirchhoff, G.   178, 192
Kirkman, T.   83
Klein, D. J.   117, 187
Kleitman, D. J.   128
König, D.   112
 (Egerváry-)König Theorem   112
Königsberg   99-100
Król, M.   81
Kronecker delta   12
Kroto, H.   76
Kuratowski, K.   67
Kuratowski's Theorem   67

Laplacian matrix   175
Lehel, J.   193
Leland, R.   180
Length of a
 binary codeword   19
 cycle   30
 partition   125
 path   2, 8
 walk   19, 73
Lesniak, L.   93, 227
Lewis, D. C.   82
Lieb, E. H.   106
Line graph   119, 188
Loop   70
Lovász, L.   115, 159, 168, 227
Lucas number   118

Mahadev, N. V. R.   227
Mahéo, M.   127
Majorization   131ff
Malekula   100
M-alternating   119
Manolopoulos, D. E.   80
Marshall, A. W.   147
Matched vertices   105
Matching   105
 number   105
 polynomial   106
Matrix-Tree Theorem   178
M-augmenting path   119
McKay, B. D.   186, 198
Medić-Sarić, M.   33
Mendeleev, D. I.   103
Menger's Theorem   52
Mercator, G.   64
Mercator projection   65

Merris, R.   149, 184, 186, 190, 203
Merris's Theorem   184
Milić, M.   122
Mohar, B.   186
Molecular orbitals   105
Morgan, A. de   69
Multigraph   18, 27, 69, 94, 210
Munkres, J. R.   92
Murty, U. S. R.   227
Mycielski, J.   42, 158

Neighbors   24
Nondeficient set of vertices   113
NP-complete   7, 23, 81, 85

O'Brien, S. C.   76
Octane   33
Olkin, I.   147
Ore, O.   86
Ore's Lemma   86
Orientable surface   71
Orientation   171
Oriented
 edge   171
 graph   171, 210
 vertex-edge incidence matrix   174
Orlin, J.   144
Overlap   25

P_4 10, 138, 158, 184
Pair group   201
Pairing Theorem   42
Palmer, E. M.   227
Paraffin   33
Part of a partition   125
Partition   125
Path
 graph   2
 in G 8
Peled, U. N.   143, 168, 227
Pendant
 edge   109
 vertex   32
Perfect
 elimination ordering   162
 graph   159, 161, 184, 189
 matching   106
 number   125
Perimeter   66, 74
Permanent   121-122
Permutation   12
Permutation matrix   12
Petersen graph   6, 84ff, 124
Planar graph   63
Plane graph   63

Index

Plummer, M. D. 227
Point of articulation 47
Polansky, O. E. 227
Pólya, G. 203, 227
Pólya's Theorem 203
Pósa, L. 88
Pósa's Theorem 88
Positive semidefinite 176
Proper
  coloring 21
  subgraph 48
Pseudograph 70
Ptolemy 64
Punctured sphere 71
Pythagoras 46

Radius 41, 144
Radosavljević, Z. 77
Radziszowski, S. P. 198
Ramsey, F. 196
Ramsey
  number 196
  property 196
Ray–Chaudhuri, D. K. 62
Read, R. C. 166, 227
Reciprocal polynomial 208
Redfield, J. H. 203
Region 63–64, 72
Regular graph 17, 114, 130
Reiner, V. 164
Residue 127
Resistance distance 187
Resonance 105
Resonance conjecture 117
Reverse–angle 37, 118
Ringel, G. 75, 79, 227
Roby, T. 149
Rouvray, D. H. 33, 227
Roy, B. 193
Ruch, E. 133
Ruch–Gutman Theorem 133, 149
Rushbrooke, G. S. 42
Ryser, H. J. 129
Ryser switch 129, 154

Sá, E. M. 164
Sachs graph 121
Sachs, H. 122ff, 227
Sachs's Theorem 123
Saclé, J.–F. 127
Sample space 114
Saturated set of vertices 113
Schmerl, J. H. 43
Schrödinger equation 80
Schur, I. 190

Schwenk, A. J. 110
SDR 114
Seinsche, D. 158
Self–complementary 8
Self–conjugate partition 136, 141
Semiregular bipartite graph 191
Separating set 47, 51
  minimum 51, 159
Shifted shape 148ff
Simeone, B. 158, 165
Simić, S. 77
Simon, H. D. 180
Simplicial vertex 42, 160
Smalley, R. 76
Spanning
  cycle 84
  directed cycle 172
  directed path 193
  forest 57
  subgraph 45
  tree 57, 59, 178
Spectral Theorem 181
Spectrum 176
Spencer, J. 227
Spialter, L. 122
Split
  graph 150
  partition 150
Sridharan, M. R. 166
Stanley, R. P. 172
Stanley's Theorem 172
Star 2
Stein, S. K. 99
Stockmeyer, L. J. 81
Subdivide an edge 39, 66
Subdivision 67
Subgraph 8
System of distinct representatives 114

Tail (of an oriented edge) 171
Term rank 118
Theta graph 101
Threshold
  Algorithm 139
  graph 135, 150ff, 184, 190
  labeling 144
  partition 135, 149ff
Toft, B. 81, 227
Topological index 33, 118
Tournament 171, 193
Trace of a partition 133
Transitive orientation 189
Transpose 18, 30, 174
Tree 31

Triangle 138
Triangulated plane graph 65
Triangulation 65
Trinajstić, N. 227
Trivial component 136
Tutte, W. T. 42, 62, 97, 115
Tutte's Theorem 115

Underlying graph 70
Unicursal graph 101
Unimodular
 congruence 177
 matrix 177
Union of graphs 23

Valence 33
Vertex 3
 covering 105
 covering number 112
 cut 47
 deleted graph 47
Vizing, V. G. 209

Wagner, K. 76
Walk 19, 73, 99
 closed 73, 101
Watkins, W. 177
Weak majorization 147
West, D. B. 227
Wheel 37, 39, 119
Whitney, H. 57, 61, 70, 92
Whitney's
 Broken Cycle Theorem 57, 194
 (planar graph) Theorem 92
Wiener, H. 33, 34, 186
Wiener Index 33, 186ff
Wilf, H. S. 36
Wolkowicz, H. 164
Woodall, D. R. 62

Young, A. 131
Young diagram 131
Youngs, J. W. 75, 79

Zalcstein, Y. 136, 144

Index of Notation

$\cong$	isomorphic	5
$\succ$	majorization	131
$\succ_w$	weak majorization	147
$\langle u_1, u_2, \ldots, u_r \rangle$	cycle of length r	30
$[u_1, u_2, \ldots, u_r]$	path of length $r-1$	8
$\lfloor x \rfloor$	greatest integer $\leq x$	17
$\lceil x \rceil$	smallest integer $\geq x$	23
$\alpha(G)$	independence number	22
$\alpha(\pi)$	partition associated with $A(\pi)$	148
A^t	transpose of matrix A	18, 174
$A^\dagger$	adjugate of matrix A	177
$a(G)$	algebraic connectivity	180
$A(G)$	adjacency matrix	11
A_n	antiregular graph	40
$A(\pi)$	above diagonal part of $F(\pi)$	147
$\beta(G)$	(vertex) covering number	112
$\beta(\pi)$	partition associated with $B(\pi)$	148
$B(\pi)$	below diagonal part of $F(\pi)$	147
$B(T)$	Balaban centric index	35
$\chi(G)$	chromatic number	23
$\chi(\Pi_p)$	max $\chi(G)$, $G \subset \Pi_p$	75
C_n	cycle on n vertices	2
$C_n H_{2n+2}$	paraffin family	33
$C(n, r)$	binomial coefficient n–choose–r	3
$\Delta(G)$	largest vertex degree	4
$\delta(G)$	smallest vertex degree	4
δ_{ik}	Kronecker–delta	12
$d_G(v)$	degree of vertex v	4
$d(v)$	degree of vertex v	4
$d(G)$	degree sequence	4
$d(u, v)$	distance from u to v	30
diam(G)	diameter of G	31

Index of Notation

$\varepsilon(G)$	edge connectivity	45
$E(G)$	edge set	3
$F(\pi)$	Ferrers diagram of partition π	131
$f(\gamma)$	trace of partition γ	133
$f_n(x)$	graph generating function	19, 199
$\gamma(G)$	genus	72
γ_n	$\gamma(K_n)$	74
$\overline{G}$	closure	86
$2G$	$G \oplus G$	23
G^c	complement	9
G^d	dual pseudograph	69
G_d	graph underlying G^d	70
$G|e$	multigraph obtained from $G-e$	27
G/e	contraction	27
$G-e$	graph obtained by deleting edge e	26
$G-S$	graph obtained by deleting $S \subset V(G)$	47
$G-v$	vertex deleted graph	10, 47
$G-Y$	graph obtained by deleting $Y \subset E(G)$	45
$G+e$	graph obtained by adding edge e	37, 118
$G \oplus H$	(disjoint) union	23
$G \vee H$	join	23
$g(n,m)$	number of graphs	19, 199
$G_{n,r}$	Chvátal graph	89
$G[W]$	subgraph induced by W	8
$\mathbf{i}(G)$	intersection number	19
K_n	complete graph	9
$K_{s,t}$	complete bipartite graph	30
$\kappa(G)$	connectivity	47
$\kappa(u,v)$	$\kappa_G(u,v)$	52
$\kappa_G(u,v)$	minimum separating set number	51
$\lambda_i(G)$	ith eigenvalue of $L(G)$	176
$\ell(\pi)$	length of partition π	125
$L(G)$	Laplacian matrix	175
$\mu(G)$	matching number	105
$m(G,r)$	number of r-matchings	106
$M(G,x)$	matching polynomial	106
$N_G(v)$	neighbors of v	24
$N_G(S)$	union over $v \in S$ of $N_G(v)$	113
$\omega(G)$	clique number	22
$o(S)$	cardinality of S	5

Index of Notation

$\pi \vdash r$	π is a partition of r	125
π^*	partition conjugate to π	133
Π_p	orientable surface of genus p	75
$\psi_G(u, v)$	max number of internally disjoint paths	52
$\psi(u, v)$	$\psi_G(u, v)$	52
P_n	path on n vertices	2
$p(G, r)$	number of proper r-colorings	21
$p(G, x)$	chromatic polynomial	29
$\tilde{p}$	induced permutation	202
per(A)	permanent of matrix A	121
Q^t	transpose of matrix Q	174
Q_k	cube graph on 2^k vertices	33
$Q(G)$	vertex-edge incidence matrix	174
$r^{(n)}$	falling factorial	22
$r(G)$	number of regions in plane graph G	64
$r(G)$	Havel–Hakimi residue	127
$r(\pi)$	Havel–Hakimi residue	127
$r(s, t)$	Ramsey number	196
$s(G)$	Laplacian spectrum	176
S_n	star on n vertices	2
S_n	permutations of $\{1, 2, \ldots, n\}$	121, 201
S_V	permutations of V	201
$S_V^{(2)}$	pair group	201
$t(G)$	spanning tree number	178ff
uv	edge $\{u, v\}$	4
$V^{(2)}$	2-element subsets of V	3
$V(G)$	vertex set	3
$V \times V$	ordered pairs of vertices	171
$x^{(n)}$	falling factorial	22, 38
$\xi(G)$	number of components	85
$\xi_o(G)$	number of odd components	115
Y_n	$n \times 1$ matrix of 1's	176
Z_n	cycle index polynomial	202

WILEY-INTERSCIENCE
SERIES IN DISCRETE MATHEMATICS AND OPTIMIZATION

ADVISORY EDITORS

RONALD L. GRAHAM

AT & T Laboratories, Florham Park, New Jersey, U.S.A.

JAN KAREL LENSTRA

*Department of Mathematics and Computer Science,
Eindhoven University of Technology, Eindhoven, The Netherlands*

AARTS AND KORST • Simulated Annealing and Boltzmann Machines: A Stochastic Approach to Combinatorial Optimization and Neural Computing
AARTS AND LENSTRA • Local Search in Combinatorial Optimization
ALON, SPENCER, AND ERDŐS • The Probabilistic Method, Second Edition
ANDERSON AND NASH • Linear Programming in Infinite-Dimensional Spaces: Theory and Application
AZENCOTT • Simulated Annealing: Parallelization Techniques
BARTHÉLEMY AND GUÉNOCHE • Trees and Proximity Representations
BAZARRA, JARVIS, AND SHERALI • Linear Programming and Network Flows
CHANDRU AND HOOKER • Optimization Methods for Logical Inference
CHONG AND ZAK • An Introduction to Optimization
COFFMAN AND LUEKER • Probabilistic Analysis of Packing and Partitioning Algorithms
COOK, CUNNINGHAM, PULLEYBLANK, AND SCHRIJVER • Combinatorial Optimization
DASKIN • Network and Discrete Location: Modes, Algorithms and Applications
DINITZ AND STINSON • Contemporary Design Theory: A Collection of Surveys
DU AND KO • Theory of Computational Complexity
ERICKSON • Introduction to Combinatorics
GLOVER, KLINGHAM, AND PHILLIPS • Network Models in Optimization and Their Practical Problems
GOLSHTEIN AND TRETYAKOV • Modified Lagrangians and Monotone Maps in Optimization
GONDRAN AND MINOUX • Graphs and Algorithms *(Translated by S. Vajdā)*
GRAHAM, ROTHSCHILD, AND SPENCER • Ramsey Theory, Second Edition
GROSS AND TUCKER • Topological Graph Theory
HALL • Combinatorial Theory, Second Edition
HOOKER • Logic-Based Methods for Optimization: Combining Optimization and Constraint Satisfaction
IMRICH AND KLAVŽAR • Product Graphs: Structure and Recognition
JANSON, LUCZAK, AND RUCINSKI • Random Graphs
JENSEN AND TOFT • Graph Coloring Problems
KAPLAN • Maxima and Minima with Applications: Practical Optimization and Duality
LAWLER, LENSTRA, RINNOOY KAN, AND SHMOYS, Editors • The Traveling Salesman Problem: A Guided Tour of Combinatorial Optimization
LAYWINE AND MULLEN • Discrete Mathematics Using Latin Squares
LEVITIN • Perturbation Theory in Mathematical Programming Applications
MAHMOUD • Evolution of Random Search Trees
MAHMOUD • Sorting: A Distribution Theory
MARTELLI • Introduction to Discrete Dynamical Systems and Chaos
MARTELLO AND TOTH • Knapsack Problems: Algorithms and Computer Implementations
McALOON AND TRETKOFF • Optimization and Computational Logic
MERRIS • Graph Theory
MINC • Nonnegative Matrices
MINOUX • Mathematical Programming: Theory and Algorithms *(Translated by S. Vajdā)*
MIRCHANDANI AND FRANCIS, Editors • Discrete Location Theory
NEMHAUSER AND WOLSEY • Integer and Combinatorial Optimization

NEMIROVSKY AND YUDIN • Problem Complexity and Method Efficiency in Optimization *(Translated by E. R. Dawson)*
PACH AND AGARWAL • Combinatorial Geometry
PLESS • Introduction to the Theory of Error-Correcting Codes, Third Edition
ROOS AND VIAL • Ph. Theory and Algorithms for Linear Optimization: An Interior Point Approach
SCHEINERMAN AND ULLMAN • Fractional Graph Theory: A Rational Approach to the Theory of Graphs
SCHRIJVER • Theory of Linear and Integer Programming
TOMESCU • Problems in Combinatorics and Graph Theory *(Translated by R. A. Melter)*
TUCKER • Applied Combinatorics, Second Edition
WOLSEY • Integer Programming
YE • Interior Point Algorithms: Theory and Analysis